G. Pólya   R.C. Read

Combinatorial Enumeration of Groups, Graphs, and Chemical Compounds

With 17 Illustrations

Springer-Verlag
New York Berlin Heidelberg
London Paris Tokyo

R.C. Read
Department of Combinatorics
  and Optimization
University of Waterloo
Waterloo, Ontario N2L, 3G1
Canada

Translated by:
Dorothee Aeppli
Division of Biometry, School of Public Health
University of Minnesota
Minneapolis, MN 55455
U.S.A.

Library of Congress Cataloging-in-Publication Data
Pólya, George
  Combinatorial enumeration of groups, graphs, and chemical compounds.
  Bibliography: p.
  1. Combinatorial enumeration problems.
I. Read, Ronald C. II. Title.
QA164.8.P65 1987   511'.62   86-31634

Printed and bound by R.R. Donnelley and Sons, Harrisonburg, Virginia.
Printed in the United States of America.

9 8 7 6 5 4 3 2 1

ISBN 0-387-96413-4 Springer-Verlag New York Berlin Heidelberg
ISBN 3-540-96413-4 Springer-Verlag Berlin Heidelberg New York

PREFACE

In 1937 there appeared a paper that was to have a profound influence on the progress of combinatorial enumeration, both in its theoretical and applied aspects. Entitled *Kombinatorische Anzahlbestimmungen für Gruppen, Graphen und chemische Verbindungen*, it was published in *Acta Mathematica*, Vol. 68, pp. 145 to 254. Its author, George Pólya, was already a mathematician of considerable stature, well-known for outstanding work in many branches of mathematics, particularly analysis.

The paper in question was unusual in that it depended almost entirely on a single theorem -- the "Hauptsatz" of Section 4 -- a theorem which gave a method for solving a general type of enumeration problem. On the face of it, this is not something that one would expect to run to over 100 pages. Yet the range of the applications of the theorem and of its ramifications was enormous, as Pólya clearly showed. In the various sections of his paper he explored many applications to the enumeration of graphs, principally trees, and of chemical isomers, using his theorem to present a comprehensive and unified treatment of problems which had previously been solved, if at all, only by ad hoc methods. In the final section he investigated the asymptotic properties of these enumerational results, bringing to bear his formidable insight as an analyst.

Pólya's paper was, and should still be, essential reading for anyone pursuing research in any field relating to enumeration, but for many English-speaking mathematicians the fact that it was written in German was an obstacle not lightly to be overcome. My own experience in this regard was no doubt similar to that of many others. I well remember how, as a raw Ph.D. student still unsure of the direction in which I wanted my research to go, I became convinced, for some reason, that Pólya's paper was one that I ought to read. Tackling such a long paper was a daunting task, for my German was rudimentary and the only copy of the paper available to me was a rather poor microfilm. But I struggled through, and was amply rewarded. Not only did the paper provide me with a powerful tool for the solution of combinatorial problems, it also revealed a whole vista of possible lines of investigation. The research path ahead was made clear to me.

Mathematical papers in German are not routinely translated into English as are those in Russian or Chinese, but in view of its obvious importance and the fact that it is as long as many books, it is surprising that Pólya's paper has not been translated before now. True, there have been rumors from time to time during the last few decades that someone, somewhere, was planning a translation, but none of these projects ever materialized. Now at last a translation has been produced, and is presented in this volume. It appears, very fittingly, on the 50th anniversary of the publication of the original paper.

Pólya's paper can truly be said to be a classic of mathematical literature -- worthwhile reading for anyone with an interest in combinatorics; but as with most classics, it is not enough to read the paper by itself. Largely as a result of Pólya's work, the subject of combinatorial enumeration has blossomed greatly during the last 50 years, and the importance of Pólya's paper can be properly appreciated only in light of these later developments. The publishers have done me the honor of asking me to supply an article to accompany the translation, giving a survey of the many different kinds of research that have stemmed from Pólya's work. In doing this I have, perforce, had to be selective -- it would be quite impossible to say something about every paper that has made use of Pólya's theorem -- but I have tried to indicate the main streams of development and to show the tremendous diversity of problems to which the theorem can be applied.

It is a great pity that Pólya did not live to see the completion of this translation. He died on September 7, 1985, having achieved the ripe old age of 97. This volume, appearing as it does on the 100th anniversary of his birth, serves as a fitting tribute to one of the most outstanding mathematicians of our time.

Ronald C. Read
University of Waterloo

CONTENTS

INTRODUCTION

1. This paper presents a continuation of work done by Cayley. Cayley has repeatedly investigated combinatorial problems regarding the determination of the number of certain trees[1]. Some of his problems lend themselves to chemical interpretation: the number of trees in question is equal to the number of certain (theoretically possible) chemical compounds.

Cayley's extensive computations have been checked and, where necessary, adjusted. Real progress has been achieved by two American chemists, Henze and Blair[2]. Not only did the two authors expand Cayley's computations, but they also improved the method and introduced more classes into the compound. Lunn and Senior[3], on the other hand, discovered independently of Cayley's problems that certain numbers of isomers are closely related to permutation groups. In the present paper, I will extend Cayley's problems in various ways, expose their relationship with the theory of permutation groups and with certain functional equations, and determine the asymptotic values of the numbers in question. The results are described in the next four chapters. More detailed summaries of these chapters are given below. Some of the results presented here in detail have been outlined before[4].

2. The combinatorial problem on permutation groups stands out for its generality and the simplicity of the solution. The following

---

[1]Cayley, 1-8.

[2]Blair and Henze, 1-6.

[3]Lunn and Senior, 1.

[4]Pólya, 1-5.

example reveals the close relationship of this problem with the first elements of combinatorics.

Suppose you have six balls with three different colors, three red, two blue, one yellow. Balls of the same color cannot be distinguished. In how many ways can you assign the six balls to the six vertices of an octahedron which moves freely in space? If the octahedron is fixed in space in such a way that the vertices are designated as upper, lower, front, back, left, and right vertex, then the number is determined by basic permutation principles as

$$\frac{6!}{3!\ 2!\ 1!} = 60.$$

The crux of the problem lies in the fact that the vertices are neither completely identifiable nor completely indistinguishable, but that those and only those among the 60 arrangements which can be transformed into each other by rotations of the octahedron, may be considered indistinguishable.

To answer the question one has to examine carefully the permutations which correspond to the 24 rotations of the octahedron. We partition these permutations into cycles and assign to each cycle of a certain order k the symbol f_k: assign f_1 to a cycle of order 1 (vertex which is invariant under rotation), f_2 to a cycle of order two (transposition), f_3 to a cycle of order three, etc. A permutation which is decomposed into the product of cycles with no common elements is represented by the product of the symbols f_i associated with the corresponding cycles. Thus the rotations of the octahedron are described by the following products:

f_1^6 : "rest" or "identity"; that is, six first order cycles.

$f_1^2 f_4$: 90° rotation with respect to a diagonal.

$f_1^2 f_2^2$: 180° rotation with respect to a diagonal.

f_2^3 : 180° rotation with respect to the line through the midpoints of two opposite edges.

f_3^2 : 120° rotation with respect to the line connecting the centers of two opposite faces.

We note these five rotation types occur with the respective frequencies

$$1 \quad 6 \quad 3 \quad 6 \quad 8.$$

Evaluate the arithmetic mean of the 24 products assigned to the 24 rotations. I will call the resulting polynomial in the four symbols f_1, f_2, f_3, f_4,

$$(f_1^6 + 6f_1^2 f_4 + 3f_1^2 f_2^2 + 6f_2^3 + 8f_3^2)/24$$

the cycle index of the permutation group which gives rise to the octahedron group of the six vertices.

The solution of the combinatorial problem is determined by the following rule: introduce

$$f_1 = x + y + z, \qquad f_2 = x^2 + y^2 + z^2$$

$$f_3 = x^3 + y^3 + z^3, \qquad f_4 = x^4 + y^4 + z^4$$

into the cycle index and expand in powers of x, y, z. The desired number is equal to the coefficient of $x^3 y^2 z$ in this expansion. It is equal to 3, which can be checked by means of a figure. The solution for the simple combinatorial problem of six distinguishable vertices discussed above follows the same rules: The only permutation compatible with distinguishable vertices is the identity, that is, the permutation group of degree 6 with cycle index f_1^6. The coefficient of $x^3 y^2 z$ in the expansion of $(x + y + z)^6$ is precisely

$$\frac{6!}{3! \, 2! \, 1!} \, .$$

Chapter 1 expands on the above introduced concept of "configurations which are equivalent with respect to a permutation group". General rules are established and some related topics are mentioned.

3. Cayley defines a tree as a geometric-combinatorial structure consisting of "vertices" and "edges". Each edge connects two vertices, an arbitrary number of edges can meet in a vertex. A tree is connected; given the number of vertices, the number of edges is the smallest number necessary to connect the vertices: that is, the number of vertices is exactly one more than the number of edges, and there are no closed paths. One distinguishes between one-edged, two-edged, three-edged, etc., vertices of the tree depending on the number of edges originating in a vertex. A one-edged vertex is also an endpoint of a tree.

An arbitrary endpoint can also be marked as "root". A tree with a root will be called a planted tree; the vertices different from the root are nodes. If no root is marked, the tree is called an unrooted or free tree. From a topological point of view, two trees with the same structure are identical; the exact definition of this and some similar, less familiar notions, will be discussed in Sections 34-35. In the sequel, we use the following notations:

t_n : number of topologically different free trees with n vertices;

T_n : number of topologically different planted trees with n nodes.

iontion

Cayley has advanced the computation of t_n and T_n to an impressive degree. The definition of t_n, ignoring the notion of root, is simpler than that of T_n. However, we will see that, from an analytical point of view, T_n is easier to handle than t_n: t_n can be derived from T_n. To evaluate T_n Cayley has established the remarkable equation

(1)
$$T_1 x + T_2 x^2 + \cdots + T_n x^n + \cdots$$
$$= x(1-x)^{-T_1}(1-x^2)^{-T_2}(1-x^3)^{-T_3} \ldots (1-x^n)^{-T_n} \ldots$$

which, considered as an identity in x, allows the successive computation of the numbers T_1, T_2, ... Extracting T_1, T_2, ... from equation (1) or determining these numbers by inspection (see Fig. 1), we find

$$T_1 = 1, \quad T_2 = 1, \quad T_3 = 2, \quad T_4 = 4, \quad T_5 = 9, \quad \ldots .$$

The roots of the trees in Fig. 1 are indicated by arrows and the nodes by circles.

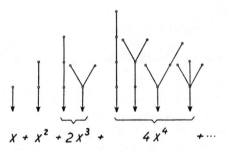

$$x + x^2 + 2x^3 + \qquad 4x^4 \qquad + \cdots$$

Figure 1

Let $t(x)$ be the generating function of the topologically different planted trees,

(2)
$$t(x) = T_1 x + T_2 x^2 + T_3 x^3 + \cdots + T_n x^n + \cdots .$$

Then Cayley's equation (1) can be interpreted as a functional equation for $t(x)$, which can be written in the following two equivalent ways (each has its own advantages):

(1')
$$t(x) = x \exp(t(x)/1 + t(x^2)/2 + \cdots + t(x^n)/n + \cdots),$$

(1")
$$t(x) = x[1 + t(x)/1! + (t^2(x) + t(x^2))/2$$
$$+ (t^3(x) + 3t(x)\, t(x^2) + t(x^3))/6 + \cdots].$$

Formula (1') is the base for asymptotic computations of T_n and t_n and (1") lends itself to generalizations. (Indeed, in the general term of the series on the right hand side of (1") we recognize the cycle index of the symmetric group of n elements.)

Cayley's equation (1) rewritten in the form of (1") serves, with proper interpretation of the group theoretical aspects, as model for any number of counts defined analogously to T_n and t_n.

Two examples of counts may suffice:

ρ_n = number of topologically different free trees consisting entirely of one- and four-edged vertices, specifically n four-edged vertices.

R_n = number of topologically different planted trees consisting entirely of one- and four-edged vertices, specifically n four-edged vertices.

The definitions of ρ_n and R_n are, from a purely geometric-combinatorial point of view, somewhat artificial. However, ρ_n is related to R_n like t_n to T_n: ρ_n will be derived from R_n, and R_n is the coefficient of x^n in the power expansion of the generating function

$$(3) \qquad r(x) = R_0 + R_1 x + R_2 x^2 + \cdots + R_n x^n + \cdots$$

satisfying the functional equation

$$(4) \qquad r(x) = 1 + x(r(x)^3 + 3r(x)r(x^2) + 2r(x^3))/6.$$

4. The chemical importance, not the geometric-combinatorial considerations, justifies the in-depth analysis of the numbers ρ_n and R_n.

A tree included in the family considered first (of size ρ_n), that is, a tree consisting of one-edged and n four-edged vertices, has exactly $2n + 2$ one-edged vertices, hence a total of $3n + 2$ vertices (see Sec. 36). By identifying the four-edged vertices with C-atoms of valence 4 and the $2n + 2$ one-edged vertices with the H-atoms of valence 1 the tree turns into the structure of a paraffin; that is, of a chemical compound with molecular formula C_nH_{2n+2}. Topologically different trees with n and $2n + 2$ vertices correspond to structurally different substances (isomers) with the common molecular formula C_nH_{2n+2}, thus

ρ_n is the number of isomers with molecular formula C_nH_{2n+2}. Similarly,

R_n is the number of isomers of molecular formula $C_nH_{2n+1}OH$ (alcohol).

The root of the tree corresponds to the OH group; the remaining $2n + 1$ endpoints correspond to H-atoms.

The interpretation of the geometric-combinatorial counts ρ_n and R_n as numbers of possible isomers, suggests that concepts in organic chemistry give rise to many analogous numbers which allow for geometric-combinatorial definition and computation. The most important numbers of this type in organic chemistry are listed below; translation into geometric-combinatorial definitions requires careful examination (see Secs. 33-36). Let

σ_n = number of stereoisomeric paraffins of molecular formula C_nH_{2n+2};

S_n = number of stereoisomeric alcohols of molecular formula $C_nH_{2n+1}OH$;

κ_n = number of structurally isomeric paraffins of molecular formula C_nH_{2n+2} without asymmetric carbon atoms;

Q_n = number of structurally isomeric alcohols of molecular formula $C_nH_{2n+1}OH$ without asymmetric carbon atoms.

The relationships between σ_n and S_n, κ_n and Q_n are similar to those of τ_n and T_n, ρ_n and R_n. The number σ_n can be calculated in terms of S_n, κ_n in terms of Q_n, while S_n and Q_n are the coefficients in the respective generating functions

(5) $s(x) = S_0 + S_1 x + S_2 x^2 + \cdots + S_n x^n + \cdots$

(6) $q(x) = Q_0 + Q_1 x + Q_2 x^2 + \cdots + Q_n x^n + \cdots$

defined by the functional equations

(7) $s(x) = 1 + x(s(x)^3 + 2s(x^2))/3$

(8) $q(x) = 1 + xq(x)q(x^2).$

Among the four functions $q(x)$, $r(x)$, $s(x)$, $t(x)$, the first, $q(x)$, has the simplest structure. Its functional equation (8) is solved by the continued fraction

(8') $q(x) = 1/(1-x/ (1-x^2/ (1-x^4/ (1-x^8/ \ldots)).$

Chapters 2 and 3 contain proofs of the claims on the numbers Q_n, R_n, S_n, T_n, κ_n, ρ_n, σ_n, τ_n, as well as discussions of a few other geometric-combinatorial, chemical-combinatorial numbers.

5. The fact that the number of isomers of homologous series increases rapidly with the number of C-atoms is well known. The methods described above allow for more precise statements and for

asymptotic evaluation of the number of isomers. Thus we are on the right track.

The combinatorial definition leads partly immediately, partly with the help of some derivations (see Secs. 36-37), to the inequalities

(9) $1 \leqslant \kappa_n \leqslant P_n \leqslant \sigma_n, \qquad P_n \leqslant T_n,$

(10) $1 \leqslant Q_n \leqslant R_n \leqslant S_n, \qquad R_n \leqslant T_n,$

(11) $P_n \leqslant R_n \leqslant nP_n, \quad \sigma_n \leqslant S_n \leqslant n\sigma_n, \quad T_n \leqslant T_n \leqslant nT_n.$

Further combinatorial considerations (Secs. 41, 43, 45) imply

(12) $$S_n \leqslant \frac{1}{n}\binom{3n}{n-1}, \quad \frac{n^{n-1}}{n!} \leqslant T_n \leqslant \frac{1}{n}\binom{2n-2}{n-1}.$$

Denote the radii of convergence of the four series $q(x)$, $r(x)$, $s(x)$, $t(x)$ by κ, ρ, σ, τ, respectively. In the limit, the inequalities (10) and (12) turn into the following inequalities between κ, ρ, σ, τ, but they are less precise than the inequalities deduced from the functional equations (1'), (4), (7), (8):

(13) $1 > \kappa > \rho > \sigma \qquad \rho > \tau$

(14) $\sigma > 4/27 \qquad \dfrac{1}{e} > \tau > \dfrac{1}{4}.$

Determination of the radii of convergence κ, ρ, σ, τ is the first step in the asymptotic evaluation of the respective combinatorial counts. The next step consists in examining the four power series $q(x)$, $r(x)$, $s(x)$, $t(x)$ on the circle of convergence. Each of the four series has exactly one isolated singular point on the circle of convergence; in fact, it lies on the positive real axis. The singular point is a pole of first order for $q(x)$, and for $r(x)$, $s(x)$, $t(x)$, it is an algebraic branch point of first order in the neighborhood of which the function is bounded. The asymptotic behavior of Q_n, R_n, S_n, T_n can now be deduced easily.

I will use the following notations:
Suppose

$$\lim_{n\to\infty} A_n/B_n = C,$$

where C is a positive real number $(0 < C < \infty)$. Then

(15) $A_n \approxeq B_n,$

which means A_n is asymptotically proportional to B_n; C is called the proportionality factor. Asymptotic equality

$$A_n \sim B_n$$

holds if A_n and B_n are asymptotically proportional with proportionality factor 1.

The analytic methods described above lead to

(16) $Q_n \approx \kappa^{-n},\ R_n \approx \rho^{-n} n^{-3/2},\ S_n \approx \sigma^{-n} n^{-3/2},\ T_n \approx \tau^{-n} n^{-3/2},$

whence

(17) $\kappa_n \approx \kappa^{-n},\ \rho_n \approx n^{-5/2} \rho^{-n},\ \sigma_n \approx \sigma^{-n} n^{-5/2},\ \tau_n \approx \tau^{-n} n^{-5/2}$.

The relationship between Q_n and κ_n is particularly simple:

(18) $Q_n \sim 2\kappa_n$.

Some generalizations of the asymptotic formulas regarding R_n and ρ_n in chemical context will emphasize the importance of these results.

The number of structurally isomeric hydrocarbons with formula $C_n H_{2n+2-2\mu}$ is asymptotically proportional to $\rho^{-n} n^{(3\mu-5)/2}$. For the paraffins μ is equal to zero, that is, the number is asymptotically proportional to ρ_n. For $\mu = 1$, the proportionality factor is 1/4.

Let X', X'', X''', ..., $X^{(\ell)}$ be mutually different radicals with valence 1. The number of structurally isomeric compounds of formula $C_n H_{2n+2-\ell}\ X'\ X''\ ...\ X^{(\ell)}$ is asymptotically proportional to $\rho^{-n} n^{(2\ell-5)/2}$. These compounds are really "ℓ-fold substituted paraffins"; the radicals have to be different from each other and from alkyls. In case of $\ell = 0$ and $\ell = 1$, the mentioned result gives the asymptotic behavior of ρ_n and R_n respectively. The proportionality factor is of the form $L\lambda^\ell$, where L and λ are independent of ℓ.

The number of isomeric homologues of benzene with formula $C_{6+n} H_{6+2n}$ is asymptotically proportional to the number of isomeric alcohols $C_n H_{2n+1} OH$ with proportionality factor $[r(\rho^5) + r(\rho)r(\rho^2)^2]/2$. Similarly, the increase in the number of isomers in other homologous series (e.g., in the series starting with naphthalene and anthazene) is asymptotically proportional to the number R_n of isomers of the alcohol series. The proportionality factor can easily be derived from the cycle index of the permutation group of the replaceable bonds of the basic compound.

6. The preceding four sections summarize only part of the content of the following four chapters; several interesting results have not been mentioned. To keep the paper within bounds, I had to forego detailed discussion of aspects I deemed less important. For this subject matter, definitions and even formal calculations and heuristic deductions seem to me often more important than complete proofs. Thus, proofs were eliminated first; in particular, in the case of several analogous propositions the proof of only one theorem is

given, the proofs of the others are left, some with hints, to the reader. Obvious steps are omitted.

Chapter 1
GROUPS

Definitions

7.   We begin by generalizing the problem which is at the root of the example in Sec. 2.   There are two types of generalizations: on the one hand the colored balls discussed in Sec. 2 have to be replaced by more complex objects, which we will call figures; on the other hand, the special permutation group of the octahedron rotations will have to be replaced by a more general permutation group.

The definitions concerning figures will be followed by those on permutation groups.   The terminology is suggestive.   Symbols will have the same meaning throughout.

8.   *Collection of figures*:   Consider a series of distinct objects ϕ', ϕ'', ..., $\phi^{(\lambda)}$, ... called figures.   The collection of these figures is the set $[\phi]$.

The figure $\phi^{(\lambda)}$ contains three categories of balls, α_λ are red, β_λ are blue, γ_λ are yellow ($\lambda = 1, 2, ...$)[1]; the figure $\phi^{(\lambda)}$ has, for short, content $(\alpha_\lambda, \beta_\lambda, \gamma_\lambda)$.

Different figures may have the same number of balls of each color.   Let $a_{k\ell m}$ denote the number of figures of content (k, ℓ, m). The power series

$$(1.1) \qquad \sum_{k=0}^{\infty} \sum_{\ell=0}^{\infty} \sum_{m=0}^{\infty} a_{k\ell m} x^k y^\ell z^m = \sum_{k,\ell,m} a_{k\ell m} x^k y^\ell z^m = f(x,y,z)$$

is the generating function of the collection $[\phi]$.

The values $a_{k\ell m}$ are supposed to be finite.   There are no assumptions on the convergence of the power series (1.1); formal expansions

---

[1]Restricting attention to three categories instead of arbitrarily many categories is not an essential limitation.

are used to describe purely algebraic manipulations with the coefficients. In the example of Sec. 2 we deal with only three different figures. The first consists of a red, the second of a blue, the third of a yellow ball, with content $(1,0,0)$, $(0,1,0)$, $(0,0,1)$, respectively. The generating function of this figure collection is

$$x + y + z.$$

The series (2) of Sec. 3, too, is a generating function; the collection of figures comprises the planted trees which are topologically different. The nodes of the rooted trees play the role of the balls in the figure; there is only one category of balls, and thus the series depends only on one variable. Figure 1 indicates how the figures (planted trees) of the same content (number of nodes) are combined in the coefficients.

9. Occasionally it is advantageous to represent the figure $\Phi^{(\lambda)}$ by a variable which we can denote by the same symbol.

Consider the series

(1.2)
$$\Phi' x^{\alpha_1} y^{\beta_1} z^{\gamma_1} + \Phi'' x^{\alpha_2} y^{\beta_2} z^{\gamma_2} + \cdots +$$
$$\Phi^{(\lambda)} x^{\alpha_\lambda} y^{\beta_\lambda} z^{\gamma_\lambda} + \cdots = \sum_{[\phi]} \Phi\, x^\alpha y^\beta z^\gamma$$

where the sum extends over the entire collection of figures; Φ denotes the general figure of content (α, β, γ) from $[\phi]$.

I will call the series (1.2) the figured power series of the collection $[\Phi]$. Setting $\phi' = \phi'' = .. = 1$ in the figured series we get the generating function (called "*counting power series*" by Pólya). Later on we will make use of the obvious relationship between the series (1.1) and (1.2).

10. *Permutation groups.* Consider a permutation group $\mathfrak{H}$ of order h and degree s.

A permutation is of type $[j_1, j_2, ..., j_s]$ if it contains j_1 cycles of order 1, j_2 cycles of order 2, ..., j_s cycles of order s. A cycle of order 1 leaves an object invariant. Obviously, the cycles are meant to have no common elements, thus,

(1.3) $1 \cdot j_1 + 2 \cdot j_2 + \cdots + s j_s = s$

is the total of permutated objects. We denote the number of permutations of type $[j_1, ..., j_s]$ of the group $\mathfrak{H}$ by $h_{j_1 j_2 \cdots j_s}$. Obviously, we have

(1.4) $\sum_{(j)} h_{j_1 j_2 \cdots j_s} = h \,,$

where the sum is over all types, i.e., all sets of non-negative integers $j_1, ..., j_s$ which satisfy equation (1.3).

Let $f_1, f_2, ..., f_s$ be independent variables. Consider the polynomial

$$(1.5) \qquad \frac{1}{h} \sum_{(j)} h_{j_1...j_s} f_1^{j_1} \cdots f_s^{j_s},$$

which is determined by the numbers $h_{j_1...j_s}$. This polynomial (1.5) is

the cycle index of $\mathfrak{H}$.[1] The cycle index is an isobaric polynomial of weight s if we assign the weight σ to f_σ ($\sigma = 1, 2, ..., s$) (see (1.3)). The coefficients in the cycle index are non-negative rational numbers of sum 1 and their lowest common denominator is h. (In Sec. 2, the cycle index of the permutation group of order 24 and degree 6 has been established.)

11. We now have to establish the relationship between the permutation group $\mathfrak{H}$ and the figure collection $[\phi]$.

We think of the objects which are interchanged by the h elements of the group $\mathfrak{H}$ as s fixed points in space. (In the example of Sec. 2, we have $s = 6$ and the points in space are the six vertices of an octahedron.) We denote the s points by $1, 2, ..., s$ and assign an arbitrary figure Φ_σ to the point denoted by σ, thus obtaining the configuration $(\phi_1, \phi_2, ..., \phi_s)$. The ϕ_σ's, $\sigma = 1, ..., s$, need not be different. Two configurations, $(\phi_1, ..., \phi_s)$ and $(\phi_1', ..., \phi_s')$, are the same if

$$\phi_1 = \phi_1', \quad \phi_2 = \phi_2', ..., \phi_s = \phi_s',$$

i.e., if the same figures of the collection $[\phi]$ are matched with the s points. The configuration $(\phi_1, \phi_2, ..., \phi_s)$ has content (k, ℓ, m) if the s figures $\phi_1, \phi_2, ..., \phi_s$ contain a total of k red, ℓ blue, and m yellow balls.

Let

$$(1.6) \qquad S = \begin{bmatrix} 1 & 2 & 3 & \cdots & s \\ i_1 & i_2 & i_3 & \cdots & i_s \end{bmatrix}$$

be a permutation of s objects; S transforms the configuration $(\phi_1, \phi_2, ..., \phi_s)$ into $(\phi_{i_1}, \phi_{i_2}, ..., \phi_{i_s})$. Two configurations are equivalent with

respect to $\mathfrak{H}$ if there exists a permutation in $\mathfrak{H}$ which converts one configuration into the other.

Each configuration is, with respect to $\mathfrak{H}$, equivalent to itself, because the identity is an element of $\mathfrak{H}$. But different configurations

---

[1] In an earlier paper [Pólya, 4], I used "Symmetrieformel" instead of "cycle index".

can be equivalent, too. The configurations which are equivalent with respect to $\mathfrak{H}$ to each other, form a transitivity system. The configurations of a transitivity system have the same content.

Let $A_{k\ell m}$ be the number of different transitivity systems of configurations with content (k,ℓ,m). In other words, $A_{k\ell m}$ is *the number of nonequivalent configurations of content $(k\ell,m)$ with respect to $\mathfrak{H}$.*

12. The problem of which the example in Sec. 2 represents a very special case can be stated as follows: *Given the collection of figures $[\phi]$, the permutation group $\mathfrak{H}$ and the content (k,ℓ,m), determine the number $A_{k\ell m}$ of nonequivalent configurations of content (k,ℓ,m) with respect to $\mathfrak{H}$.* List the desired numbers $A_{k\ell m}$ in the power series

$$\sum_{k=0}^{\infty} \sum_{\ell=0}^{\infty} \sum_{m=0}^{\infty} A_{k\ell m} x^k y^\ell z^m = \sum_{k,\ell,m} A_{k\ell m} x^k y^\ell z^m = F(x,y,z).$$

That is, $F(x,y,z)$ is the generating function of the number of nonequivalent configurations. The solution of our problem consists in expressing the generating function $F(x,y,z)$ in terms of the generating function $f(x,y,z)$ of the collection of figures and the cycle index of the permutation group $\mathfrak{H}$.

Preliminary Problems

13. First, consider the special case of Sec. 12 with $\mathfrak{H}$ being the symmetric group $\mathfrak{S}_s$ of degree s. In this case a configuration is equivalent to any other configuration which can be obtained by one of the $s!$ permutations. Thus, it does not matter in what sequence the s figures of a configuration are assigned to the s points in space. Only the figures involved in a configuration matter. In this special case where $\mathfrak{H} = \mathfrak{S}_s$, $A_{k\ell m}$ denotes the number of combinations with repetitions of s figures with content (k,ℓ,m). Let $F_s(k,\ell,m)$ be the corresponding generating function. We will determine all the F_i's simultaneously.

In the product

$$(1 + u\phi' x^{\alpha_1} y^{\beta_1} z^{\gamma_1} + u^2 \phi' \phi' x^{2\alpha_1} y^{2\beta_1} z^{2\gamma_1} + \ldots)$$

$$(1 + u\phi'' x^{\alpha_2} y^{\beta_2} z^{\gamma_2} + u^2 \phi'' \phi'' x^{2\alpha_2} y^{2\beta_2} z^{2\gamma_2} + \ldots)$$

(1.8)

$$\cdots \cdots$$

$$= \prod_{[\phi]} (1 + u\phi \, x^{\alpha} y^{\beta} z^{\gamma} + u^2 \phi\phi \, x^{2\alpha} y^{2\beta} z^{2\gamma} + \ldots),$$

each combination with repetition of s figures ϕ_1, ..., ϕ_s of content (k,ℓ,m) is represented by a term

$$u^s \, \phi_1 \cdots \phi_s \, x^k \, y^\ell \, z^m.$$

Therefore, setting

(1.9) $\qquad \phi' = \phi'' = \ldots = 1$

in the product (1.8) we find the number $A_{k\ell m}$ to be the coefficient of u^s x^k y^ℓ z^m and the coefficient of u^s is the generating function $F_s(x,y,z)$. The product (1.8) can be rewritten in the form

(1.10)
$$\prod_{[\Phi]} (1 - u\Phi \, x^\alpha \, y^\beta \, z^\gamma)^{-1}$$

$$= \exp\left[-\sum_{[\Phi]} \log(1 - u\phi \, x^\alpha \, y^\beta \, z^\gamma\right]$$

$$= \exp\left[\frac{u}{1} \sum_{[\Phi]} \Phi \, x^\alpha \, y^\beta \, z^\gamma + \frac{u^2}{2}\sum \Phi^2 \, x^{2\alpha} \, y^{2\beta} \, z^{2\gamma} + \cdots\right].$$

The first term of the last expression is the figured power series (1.2) of the collection of figures $[\Phi]$. Thus, under the condition (1.9), (1.8) and (1.10) respectively lead to

(1.11)
$$1 + uF_1(x,y,z) + u^2F_2(x,y,z) + \cdots + u^sF_s(x,y,z) + \cdots$$

$$= \prod_{k=0}^{\infty} \prod_{\ell=0}^{\infty} \prod_{m=0}^{\infty} (1 - ux^k \, y^\ell \, z^m)^{-a_{k\ell m}}$$

$$= \exp\left[\frac{u}{1} \, f(x,y,z) + \frac{u^2}{2} \, f(x^2,y^2,z^2) + \frac{u^3}{3} f(x^3,y^3,z^3) + \cdots\right]$$

$$= e^{uf(x,y,z)} \cdot e^{(u^2/2)f(x^2,y^2,z^2)} \cdots$$

$$= \sum_{j_1=0}^{\infty} \frac{u^{j_1}f(x,y,z)^{j_1}}{j_1! \, 1^{j_1}} \sum_{j_2=0}^{\infty} \frac{u^{2j_2}f(x^2,y^2,z^2)^{j_2}}{j_2! \, 2^{j_2}} \cdot$$

$$\cdot \sum_{j_3=0}^{\infty} \frac{u^{3j_3}f(x^3,y^3,z^3)^{j_3}}{j_3! \, 3^{j_3}} \cdots .$$

The third line of (1.10) becomes the third line of (1.11) because for (1.9)

$$(\phi^{(\lambda)})^2 = (\phi^{(\lambda)})^3 = \cdots = 1$$

holds as well. Comparison of the coefficients of u^s in the first and final line of (1.11) yields

$$F_s(x,y,z) = \frac{1}{s!} \sum_{(j)} \frac{s!}{j_1! \, 1^{j_1} \, j_2! \, 2^{j_2} \, \cdots \, j_s! \, s^{j_s}} \cdot$$

(1.12)

$$\cdot f(x,y,z)^{j_1} f(x^2,y^2,z^2)^{j_2} \cdots f(x^s,y^s,z^s)^{j_s} \, ,$$

where in the notation of Sec. 10 summation over (j) means summation over all types of permutations of s objects.

14. A minor change in the computation above yields the numbers of combinations without repetitions of s figures from $[\Phi]$ with content (k,l,m). We call this number B_{klm} and define the generating function

$$\sum_{k,l,m} B_{klm} x^k \, y^l \, z^m = G_s(x,y,z).$$

Expanding the product

$$(1 + u\Phi' \, x^{\alpha_1} y^{\beta_1} z^{\gamma_1})(1 + u\Phi'' \, x^{\alpha_2} y^{\beta_2} z^{\gamma_2}) \, \cdots$$

(1.13)

$$= \prod_{[\Phi]} (1 + u\phi \, x^\alpha y^\beta z^\gamma)$$

in a power series we find each combination without repetitions of s figures with content (k,l,m) represented by a term

$$u^s \, \Phi_1 \, \Phi_2 \, \cdots \, \Phi_s \, x^k \, y^l \, z^m.$$

Introducing relation (1.9) in (1.13), we get

$$1 + uG_1(x,y,z) + u^2 G_2(x,y,z) + \cdots + u^s G_s(x,y,z) + \cdots$$

$$= \prod_{k=0}^{\infty} \prod_{l=0}^{\infty} \prod_{m=0}^{\infty} (1 + ux^k y^l z^m)^{a_{klm}}$$

$$= \exp\left[\frac{u}{1} f(x,y,z) - \frac{u^2}{2} f(x^2,y^2,z^2) + \frac{u^3}{3} f(x^3,y^3,z^3) - \cdots\right],$$

and by comparing coefficients

$$G_s(x,y,z) = \frac{1}{s!} \sum_{(j)} \frac{s!(-1)^{j_2+j_4+\cdots}}{j_1! \, 1^{j_1} \, j_2! \, 2^{j_2} \, \cdots \, j_s! \, s^{j_s}}$$

(1.14)

$$\cdot f(x,y,z)^{j_1} f(x^2,y^2,z^2)^{j_2} \cdots f(x^s,y^s,z^s)^{j_s} \, .$$

15. The solution of the problem in the special case of the symmetric group, $\mathfrak{H} = \mathfrak{S}_s$, allows the solution of another special case, namely for $\mathfrak{H} = \mathfrak{A}_s$, the alternating group of degree s. Consider two

configurations, C and C' of s figures each from $[\Phi]$. Under what conditions are C and C' equivalent?

It is necessary that C and C' are equivalent with respect to $\mathfrak{S}_s$, that is, C and C' have to contain the same combination (with repetitions) of s figures.

In one case this condition is also sufficient: if a figure appears twice in the combination which is common to C and C' we can add the transposition of the two points in C to which the same figure is attached to the permutation which transforms C into C'. Thus, we can force the transformation of C into C' to be an even permutation. We conclude that *combinations with at least one repetition of a figure give rise to one single transitivity system.*

It is easy to see that a combination with no repetitions gives rise to exactly two transitivity systems with respect to $\mathbf{A}_s$. Summarizing the results, we have the rule: the number of different transitivity systems of configurations with respect to $\mathbf{A}_s$ is the sum of the respective numbers of combinations with and without repetitions. Therefore, the generating function of the permutations which are nonequivalent with respect to $\mathbf{A}_s$ is

(1.15) $F_s(x,y,z) + G_s(x,y,z).$

16.   *The main theorem.*   In order to combine the results for $\mathfrak{H} = \mathfrak{S}_s$ and $\mathfrak{H} = \mathbf{A}_s$ into one expression we recall[1] that

$$\frac{s!}{j_1!\, j_2!\, 2^{j_2} \dots j_s!\, s^{j_s}}$$

is the number of permutations of s objects of type $[j_1, \dots, j_s]$. In the terminology of Sec. 10 the cycle index of the symmetric group $\mathfrak{S}_s$ is

(1.16) $\dfrac{1}{s!} \sum\limits_{(j)} \dfrac{s!}{j_1!\, 1^{j_1} \dots j_s!\, s^{j_s}} f_1^{j_1} f_2^{j_2} \dots f_s^{j_s},$

the cycle index of the alternating group $\mathbf{A}_s$ is

(1.17) $\dfrac{1}{s!} \sum\limits_{(j)} \dfrac{s![1 + (-1)^{j_2 + j_4 + \cdots}]}{j_1!\, 1^{j_1}\, j_2!\, 2^{j_2} \dots j_s!\, s^{j_s}} f_1^{j_1} f_2^{j_2} \dots f_s^{j_s}.$

We note that (1.12) relates to the cycle index (1.16) like (1.15) (taking (1.12) and (1.14) into account) to (1.17). The following definitions allow us to state the rules on the construction of the generating functions in a unified way. To introduce the functions $f(x)$, $f(x,y)$ into the cycle index means putting

---

[1] Cf. e.g. Serret, *Cours d'algèbre supérieure*, 3rd ed. (Paris 1866), Vol. 2, pp. 235-236.

(1.18) $f_1 = f(x), \quad f_2 = f(x^2), \quad f_3 = f(x^3), \; ...$

and

$$f_1 = f(x,y), \quad f_2 = f(x^2,y^2), \quad f_3 = f(x^3,y^3), \; ...$$

respectively; the generalization to functions of more variables is obvious. With this convention the results for the two special cases ($\mathfrak{H} = \mathfrak{S}_s$ and $\mathfrak{H} = \mathfrak{A}_s$) are summed up in the following main theorem.

Theorem. *The generating function for the configurations [Φ] which are nonequivalent with respect to $\mathfrak{H}$ is obtained by substituting the generating function of [Φ] in the cycle index of $\mathfrak{H}$.*

In the sequel we will see that the proposition holds for an arbitrary permutation group and we will refer to it as the theorem or the main theorem.

17. The theorem certainly holds in the special case in which the permutation group of degree s has order 1, that is, consists of the identity. Two nonidentical configurations are thus nonequivalent and the cycle index is f_1^s. In the usual terminology the solution is well known. It is a special case in a more general problem stated in terms of $s = 3$ which, however, is a nonessential restriction. *Let f, g, h denote the generating functions of three collections of figures, [Φ], [ψ], and [X]. Determine the generating function of the triple of figures (ϕ, ψ, X) where ϕ, ψ, X exhaust the respective collections [Φ], [ψ], [X] independently of each other.*

The generating function of the triple (ϕ, ψ, X) is understood to be the power series in the three variables x, y, z in which the coefficient of $x^k y^l z^m$ is equal to the number of triples whose three figures ϕ, ψ, X have content (k, l, m); that is, which contain k red, l blue, and m yellow balls. The numbers of balls of the three types are α, β, γ in ϕ, as before, α', β', γ' in ψ, $\alpha'', \beta'', \gamma''$ in X. Each triple (ϕ, ψ, X) is presented by the product $\phi \psi X$. The figured power series of the products is equal to the product of the figured power series,

$$\sum_{[\phi]} \sum_{[\psi]} \sum_{[X]} \phi \psi X \, x^{\alpha+\alpha'+\alpha''} \, y^{\beta+\beta'+\beta''} \, z^{\gamma+\gamma'+\gamma''}$$

$$= \sum_{[\phi]} \phi \, x^\alpha y^\beta z^\gamma \sum_{[\psi]} \psi \, x^{\alpha'} y^{\beta'} z^{\gamma'} \sum_{[X]} X \, x^{\alpha''} y^{\beta''} z^{\gamma''}.$$

Setting all the variables $\phi, \phi', ..., \psi, \psi', ..., X, X', ...$ equal to 1 we find that the generating function of the triple is equal to the product of the generating functions of $\phi, \psi,$ and X. This relationship indicates that the restriction of the number of factors to three is no restriction at all. It is an elementary principle, known since Euler's times, which I may state as follows: If the elements of an ordered set can be chosen independently, then the generating function of the set is equal to the product of the generating function of the individual elements.

18. The solution of the following problem illustrates further what the formulas (1.12) and (1.14) have in common.

Let the generating function (1.1) *of the collection of figures* [Φ] *and the type* [j_1, ..., j_s] *of the permutation* (1.6) *be given. Determine the generating function of the configurations* (ϕ_1, ..., ϕ_s) *of s figures from* [Φ] *which remain invariant under the permutation* (1.6).

Let $X_{k\ell m}(S)$ be the number of those configurations with content (k,ℓ,m) which remain invariant under the permutation S in (1.6). Thus, the generating function of interest is

$$\sum_{k,\ell,m} X_{k\ell m}(S)\, x^k y^\ell z^m.$$

The configuration (Φ_1, ..., Φ_s) is invariant under (1.6) if and only if

(1.19) $\Phi_{i_1} = \Phi_1,\; \Phi_{i_2} = \Phi_2,\; ...,\; \Phi_{i_s} = \Phi_s$.

Let $(a,b,c, ..., k,\ell)$ be a cycle contained in the permutation (1.6), and λ denote its length (order). If (1.6) leaves (Φ_1, ..., Φ_s) invariant, then certain λ equalities in (1.19) imply

$$\Phi_a = \Phi_b = \Phi_c = \cdots = \Phi_k = \Phi_\ell.$$

This means, the figures which belong to the same cycle of (1.6) have to be equal. In each cycle one figure from [Φ] can be arbitrarily chosen.

Therefore, a configuration which is invariant under the permutation (1.6) can be considered a set of $j_1 + j_2 + \cdots + j_\lambda + \cdots + j_s$ cycles. The figures within a cycle are identical; each cycle can be represented by an arbitrarily chosen figure. If the content of a figure in a certain cycle of length λ is (k,ℓ,m), the total content of the figures in that cycle is $(\lambda k, \lambda\ell, \lambda m)$. Hence the generating function corresponding to this cycle is

$$f(x^\lambda, y^\lambda, z^\lambda).$$

The desired generating function of the S-invariant configurations is, according to the principle given at the end of Sec. 17, a product of $j_1 + \cdots + j_s$ factors:

(1.20) $\sum_{k,\ell,m} X_{k\ell m}(S)\, x^k\, y^\ell\, z^m$

$$= f(x,y,z)^{j_1} f(x^2,y^2,z^2)^{j_2} \,...\, f(x^s,y^s,z^s)^{j_s} .$$

Determination of the Number of Nonequivalent Configurations for an Arbitrary Permutation Group

19. To solve the general problem of Sec. 12 we consider the triple of numbers k, ℓ, m and the set of all configurations of content (k,ℓ,m) (exactly k red, ℓ blue, and m yellow balls). Let C denote an arbitrary configuration and $\mathfrak{G}$ be the subgroup of permutations of $\mathfrak{H}$ that leave C invariant. There always exists such a permutation, namely the identity. Let g be the order of $\mathfrak{G}$. The number of different configurations into which C can be transformed by the permutations of $\mathfrak{H}$, that is, to which C is equivalent with respect to $\mathfrak{H}$, is h/g. Each of the h/g configurations is invariant under exactly g permutations of $\mathfrak{H}$, specifically, under the permutations of a subgroup which is conjugate to $\mathfrak{G}$ in $\mathfrak{H}$. Hence, each configuration which is equivalent to C is included in exactly g terms of the sum

$$(1.21) \qquad X_{k\ell m}(S_1) + X_{k\ell m}(S_2) + \cdots + X_{k\ell m}(S_h)$$

(notation of Sec. 18). It contributes, thus, g units to the sum. Since the number of configurations which are equivalent to C with respect to $\mathfrak{H}$ is h/g, the class of configurations which are equivalent to C, i.e., the transitivity system determined by C, contributes

$$(h/g) \cdot g = h$$

units to the sum (1.21). All the different transitivity systems of configurations which are equivalent to C with respect to $\mathfrak{H}$ contribute the same amount h, thus

$$(1.22) \qquad X_{k\ell m}(S_1) + X_{k\ell m}(S_2) + \cdots + X_{k\ell m}(S_h) = h\, A_{k\ell m}.$$

The desired generating function (1.7) results from (1.22) and (1.20):

$$F(x,y,z) = \sum_{k,\ell,m} (X_{k\ell m}(S_1) + \cdots + X_{k\ell m}(S_h))x^k\, y^\ell\, z^m/h$$

$$= (1/h)\sum_{(\mathfrak{H})} \sum_{k,\ell,m} X_{k\ell m}(S)x^k\, y^\ell\, z^m,$$

$$(1.23) \qquad F(x,y,z) = (1/h) \sum_{(\mathfrak{H})} f(x,y,z)^{j_1} f(x^2,y^2,z^2)^{j_2} \ldots f(x^s,y^s,z^s)^{j_s},$$

where the sum over $(\mathfrak{H})$ means summing over all h permutations S of the group $\mathfrak{H}$. Combining the permutations of the same type $[j_1, \ldots, j_s]$ we can rewrite formula (1.23):

$$(1.24) \qquad F(x,y,z) = (1/h) \sum_{(j)} h_{j_1 \cdots j_s} f(x,y,z)^{j_1} f(x^2,y^2,z^2)^{j_2} \ldots f(x^s,y^s,z^s)^{j_s},$$

where $h_{j_1,\ldots,j_s}$ is the number of permutations of type $[j_1, \ldots, j_s]$, as defined in Sec. 10. Recalling the definition of the cycle index (1.5) of $\mathfrak{H}$, we recognize the general theorem, which has been stated at the end of Sec. 16 on the basis of two special cases.

20. Only very elementary theorems of group theory have entered into the derivations in Sec. 19. A proof requiring more familiarity with group theory follows. Neither representation theory nor other notions introduced in this section will be used elsewhere in this paper.

Each element of $\mathfrak{H}$ effects an interchange of the s points and thus a permutation of the configurations $(\phi_1, \ldots, \phi_s)$ of content (k, ℓ, m). These permutations form a representation $\mathfrak{D}_{k\ell m}$ of the group $\mathfrak{H}$. The representation $\mathfrak{D}_{k\ell m}$ is, like $\mathfrak{H}$, a permutation group: $\mathfrak{H}$ interchanges points, $\mathfrak{D}_{k\ell m}$ interchanges configurations attached to the s points acted upon by $\mathfrak{H}$. The quantity $X_{k\ell m}(S)$ defined in Sec. 18 and determined by (1.20) is the character of the permutation in $\mathfrak{D}_{k\ell m}$ which is assigned to S of $\mathfrak{H}$. By definition in Sec. 11, the number $A_{k\ell m}$ is the number of different transitivity systems of the permutation group $\mathfrak{D}_{k\ell m}$. This number is, according to a well-known theorem,[1] equal to the arithmetic mean of the characters $X_{k\ell m}(S)$ of the permutation group $\mathfrak{D}_{k\ell m}$. To complete the proof we note that $A_{k\ell m}$ in (1.22) is equal to the arithmetic mean of the characters of $\mathfrak{D}_{k\ell m}$ even if $\mathfrak{D}_{k\ell m}$ is not a faithful but a reduced representation of $\mathfrak{H}$, in which case the order of $\mathfrak{D}_{k\ell m}$ is not h but a divisor of h.

These considerations lead to the following proposition: *Let $\mathfrak{H}$ denote a permutation group, S be a permutation of $\mathfrak{H}$ of type $[j_1, \ldots, j_s]$, $f(x,y,z)$ be an arbitrary power series in x, y, z with non-negative integer coefficients, and k, ℓ, m be a triple of natural numbers. Then the coefficient of $x^k y^\ell z^m$ in the expansion (1.20) is the character of S in a representation which is specified by $f(x,y,z)$ and k, ℓ, m.* Professor Schur communicated to me a proof of this proposition based on well-known theorems in representation theory.

The polynomial (1.5) which I called cycle index is, if $\mathfrak{H}$ is the symmetric group, equal to the principal character of $\mathfrak{H}$ in representation theory. Professor Schur informed me that the cycle index of an arbitrary permutation group being really a subgroup of a symmetric group is of importance for the representation of this symmetric group.[2] We will, however, not expand on the relationship between representation theory and our subject.

---

[1] See, e.g., A. Speiser, *Theory of Finite Groups*, 2nd Ed. (Berlin, 1927), pg. 120, Theorem 2.

[2] I. Schur, Darstellungstheorie der Gruppen, Lecture Notes, Swiss Federal Institute of Technology, E. Stiefel, ed. (Zürich, 1936) (Cf. pp. 59-60).

Professor Schur also made me aware of a consideration by Frobenius[1] which is closely related to the argument given in Sec. 19.

Special Cases

21. *Special Permutation Groups.* The following well-known special permutation groups, all of degree s (i.e., s objects are permuted) will appear in the applications of the theorem of Sec. 16:

$\mathcal{S}_s$: the symmetric group of s objects, of order $s!$;

$\mathbf{A}_s$: the alternating group of s objects which consists of the even permutations; it is of order $s!/2$;

$\mathbf{Z}_s$: the cyclic group of order and degree s, generated by cyclic permutations of s objects;

$\mathbf{D}_s$: the dihedral group of order $2s$ containing the permutations which coincide with the $2s$ deck transformations of the regular polygon with s vertices (s-gon);

$\mathbf{E}_s$: the trivial permutation group of degree s and order 1, consisting of the identity.

The cycle index plays an essential role in the theorem of Sec. 16. The cycle indices for $\mathcal{S}_s$ and $\mathbf{A}_s$ are given in (1.16) and (1.17), respectively. For small s, the cycle indices are

$(\mathcal{S}_1)$ $\qquad\qquad\qquad\qquad f_1$

$(\mathcal{S}_2)$ $\qquad\qquad\qquad (f_1^2 + f_2)/2$

$(\mathcal{S}_3)$ $\qquad\qquad\quad (f_1^3 + 3f_1 f_2 + 2f_3)/6$

$(\mathcal{S}_4)$ $\qquad\quad (f_1^4 + 6f_1^2 f_2 + 3f_2^2 + 8f_1 f_3 + 6f_4)/24$

$(\mathbf{A}_2)$ $\qquad\qquad\qquad\qquad f_1^2$

$(\mathbf{A}_3)$ $\qquad\qquad\qquad (f_1^3 + 2f_3)/3$

$(\mathbf{A}_4)$ $\qquad\qquad (f_1^4 + 3f_2^2 + 8f_1 f_3)/12.$

The cycle indices of $\mathbf{E}_s$, $\mathbf{Z}_s$, $\mathbf{D}_s$ are

$(\mathbf{E}_s)$ $\qquad\qquad\qquad\qquad f_1^s$,

---

[1]Sitzungsbericht der Akademie Berlin (1904), pp. 558-571, cf. Sec. 1.

$(\mathfrak{Z}_s)$ $\qquad \sum\limits_{k|s} \varphi(k) f_k^{s/k}/s$,

$(\mathfrak{D}_s)$ $\qquad \sum\limits_{k|s} \varphi(k) f_k^{s/k}/2s +$ $\begin{cases} f_1 f_2^{\sigma-1}/2 & \text{for } s = 2\sigma - 1 \\[2mm] (f_1^2 f_2^{\sigma-1} + f_2^\sigma)/4 & \text{for } s = 2\sigma. \end{cases}$

The cycle index of $\mathfrak{E}_s$ is obvious, for $\mathfrak{Z}_s$ and $\mathfrak{D}_s$ it is easily derived. For $\mathfrak{Z}_s$ and $\mathfrak{D}_s$ summation extends over all divisors of s. The special cases for the smallest values of s are reflected in the above table, since

$$\mathfrak{E}_1 = \mathfrak{S}_1, \quad \mathfrak{E}_2 = \mathfrak{A}_2, \quad \mathfrak{Z}_2 = \mathfrak{S}_2, \quad \mathfrak{Z}_3 = \mathfrak{A}_3, \quad \mathfrak{D}_3 = \mathfrak{S}_3.$$

22. *Special collections of figures.* Two special collections of figures deserve mentioning. They have been discussed in the context of arbitrary permutation groups in the literature.

(a) The collection of figures contains n elements. Each figure contains exactly one ball. Two different figures contain balls of different color. In short, the collection contains n balls of different colors. The generating function is

$$x_1 + x_2 + \cdots + x_n.$$

The problem of Sec. 12 can be stated in this special situation as follows: Let $\mathfrak{H}$ be an arbitrary permutation group of degree s and $k_1, k_2, \ldots, k_n$ denote n non-negative integers whose sum is s. How many nonequivalent ways modulo $\mathfrak{H}$ are there to place k_1 balls of the first, k_2 balls of the second, ..., k_n balls of the n-th color in s slots? According to Sec. 16 the solution is established by introducing $f_m = x_1^m + \cdots + x_n^m$ into the cycle index of $\mathfrak{H}$ and expanding the homogeneous polynomial of degree s. The desired number is equal to the coefficient of

$$x_1^{k_1} x_2^{k_2} \ldots x_n^{k_n}.$$

(The problem of Sec. 2 is a special case hereof.)

Lunn and Senior (see References) have dealt with this problem in a slightly different formulation. They recognized its chemical importance (see Sec. 56); their solution looks quite different from the one presented here. Lunn and Senior's solution can be considered as a special computational scheme. Since f_m is equal to the sum of the m-th powers of the variables $x_1, \ldots, x_n$, classical formulas on symmetric functions allow further inferences. Details might be discussed somewhere else.

(b) There is only one type of ball and for a given number k there is exactly one figure with k balls. In this case the generating function of the collection $[\phi]$ is

(1.25) $1 + x + x^2 + \cdots + x^k + \cdots = 1/(1-x).$

The configurations of content k which can be composed of the figures of this collection are arrangements

$$(k_1, ..., k_s)$$

of s non-negative integers $k_1, ..., k_s$ which add up to k. By assigning a variable to each point (the variable u_σ to the point σ, $\sigma = 1, ..., s$) we can describe the configuration by means of a product of powers

(1.26) $u_1^{k_1} u_2^{k_2} ... u_s^{k_s}$

of degree

$$k_1 + k_2 + \cdots + k_s = k,$$

and the permutation group $\mathfrak{H}$ is a substitution group of the s unknowns $u_1, u_2, ..., u_s$. The problem of Sec. 12 can be reformulated as follows: In how many transitivity systems can the power products (1.26) be decomposed with respect to $\mathfrak{H}$? Two power products are part of the same transitivity system if and only if there is a permutation in $\mathfrak{H}$ transforming one into the other. The sum of the products (1.26) which belong to the same transitivity system is invariant under $\mathfrak{H}$. It is easy to see that the desired number is the number of linearly independent rational entire homogenous absolute invariants of degree k of the group $\mathfrak{H}$.

According to the main theorem (Sec. 16) this number is equal to the coefficient of x^k in the expansion of the function of x which obtains by substituting (1.25) in the cycle index of $\mathfrak{H}$ or by specializing $f(x,y,z)$ in (1.23) to (1.25). The function is

(1.27) $\dfrac{1}{h} \displaystyle\sum_{(\mathfrak{H})} \dfrac{1}{(1-x)^{j_1}(1-x^2)^{j_2}...(1-x^s)^{j_s}} = \dfrac{1}{h} \displaystyle\sum_{(\mathfrak{H})} \dfrac{1}{|E - xS|}$

where on both sides the sum is over all permutations S of $\mathfrak{H}$, as in (1.23); S on the right hand side is a matrix of s rows and s columns (s of its elements are 1, $s^2 - s$ are equal to 0); E denotes the identity matrix; the denominator is the determinant $|E - xS|$ (essentially the characteristic polynomial).

We have shown that the number of linearly independent invariants of degree k under the permutation group $\mathfrak{H}$ is equal to the coefficient of x^k in the Maclaurin expansion of (1.27). This represents an important special case of a proposition by Th. Molien.[1]

---

[1]Sitzungsbericht der Akademie Berlin (1897), pp. 1152-1156. See formula (12) which holds for arbitrary finite groups of linear substitutions, not only for permutation groups. I'm obliged to Prof. Schur for this reference.

23. *Corollaries.* Many very special cases of the problem stated in Sec. 12 and solved by the main theorem appear isolated in the literature. A slightly more general special case is the case of the cyclic permutations with repetitions. It arises from the combination of the cyclic group $\mathbb{Z}_s$ with the collection of figures described in Sec. 22(a). The results discussed in the literature follow from the main theorem.[1] By combining the symmetric and alternating groups with the collection of figures of Sec. 22(a), we recover by means of the main theorem the classical formulas for symmetric functions, thus gaining further support for the approach.

In the following applications we will repeatedly encounter the special cases examined in Secs. 13 - 15 regarding the symmetric and alternating groups $\mathcal{S}_s$ and $\mathcal{A}_s$. The three polynomials in $f_1, \ldots, f_s$, derived from the cycle indices F_s and $F_s + G_s$ of these groups,

$$F_s,$$

$$G_s = (F_s + G_s) - F_s,$$

$$F_s - G_s = 2F_s - (F_s + G_s),$$

have, as we have seen, the following properties: If the generating function of [Φ] is introduced into these polynomials according to Sec. 16, then the generating function for the

combinations of s arbitrary figures,

combinations of s mutually different figures,

combinations of s not mutually different figures,

respectively, obtains. The cycle index F_s has been given for small s in Sec. 21. Below the indices G_s, $F_s - G_s$ are listed, with symbolic emphasis on the derivation from the two cycle indices:

$(\mathcal{A}_2 - \mathcal{S}_2)$	$(f_1^2 - f_2)/2$
$(\mathcal{A}_3 - \mathcal{S}_3)$	$(f_1^3 - 3f_1f_2 + 2f_3)/6$
$(\mathcal{A}_4 - \mathcal{S}_4)$	$(f_1^4 - 6f_1^2f_2 + 3f_2^2 + 8f_1f_3 - 6f_4)/24$
$(2\mathcal{S}_2 - \mathcal{A}_2)$	f_2
$(2\mathcal{S}_3 - \mathcal{A}_3)$	f_1f_2
$(2\mathcal{S}_4 - \mathcal{A}_4)$	$(f_1^2f_2 + f_4)/2.$

[1] E. Jablonski, *Journ. des mathématiques pures et appliquées* (4) 1892, pp. 331.349.

The combinatorial interpretation (or the computations of Sec. 14 regardless of combinatorial considerations) implies the following useful result: Substituting a power series with non-negative integer coefficients in the difference of the cycle indices of A_s and $\mathcal{S}_s$, we get a power series with non-negative integer coefficients.

Generalization

24. We point out a generalization of the problem of Sec. 12 which will not be taken up again (except in Sec. 65) but which might be useful in related questions.

Let $\mathfrak{H}$ be an intransitive permutation group of degree $s + t$; the elements permuted by $\mathfrak{H}$ decompose into two classes, of s and t elements, respectively. The two classes are closed with respect to $\mathfrak{H}$, that is, there is no permutation in $\mathfrak{H}$ which substitutes an element of one class with one from the other class. (The generalization from two to n classes is obvious.) Imagine the $s + t$ elements upon which $\mathfrak{H}$ operates as points in space; s figures from the collection $[\Phi]$ are assigned to the s points of the first class, t figures from $[\Psi]$ are assigned to the t points of the second class. The resulting configuration is

(1.28) $(\phi_1, \phi_2, ..., \phi_s; \psi_1, \psi_2, ..., \psi_t)$.

How many configurations of type (1.28) and content (k, ℓ, m) are there which are nonequivalent with respect to $\mathfrak{H}$?

Let S be an arbitrary permutation of $\mathfrak{H}$. A cycle involves points of one class only. Let S be of type

$$[j_1 + k_1, j_2 + k_2, ..., j_m + k_m, ...]$$

where j_m of the $j_m + k_m$ cycles of length m in S refer to points of the first class, involving figures from $[\Phi]$, and k_m to points of the second class, with figures from $[\psi]$. Thus, we have

$$1j_1 + 2j_2 + \cdots + sj_s = s,$$

$$1k_1 + 2k_2 + \cdots + tk_t = t.$$

Consider the following polynomial in $s + t$ variables $f_1, f_2, ..., f_s, g_1, g_2, ..., g_t$,

(1.29) $\dfrac{1}{h} \underset{(\mathfrak{H})}{\Sigma} f_1^{j_1} f_2^{j_2} ... f_s^{j_s} g_1^{k_1} g_2^{k_2} ... g_t^{k_t}$,

where summation is over all permutations S of $\mathfrak{H}$. Let $f(x,y,z)$ and $g(x,y,z)$ denote the generating functions of $[\Phi]$ and $[\Psi]$, respectively.

The desired number is the coefficient of $x^k y^\ell z^m$ in the expansion of (1.29) with

$$f_n = f(x^n, y^n, z^n), \qquad g_n = g(x^n, y^n, z^n), \qquad n = 1, 2, 3, \dots .$$

Relations Between Cycle Index and Permutation Group

25. The property of the cycle index given in Sec. 16 uniquely determines the cycle index. More accurately: *Let* $f(x_1, x_2, \dots)$ *denote the generating function of an arbitrary collection of figures* [ϕ]. *Suppose that the following relationship holds between the polynomial* $\Psi(f_1, f_2, \dots, f_s)$ *in the* s *variables* $f_1, f_2, \dots, f_s$ *and a given permutation group* $\mathfrak{H}$ *of degree* s: *If the variables*

$$(1.30) \qquad f_1 = f(x_1, x_2, \dots), \, f_2 = f(x_1^2, x_2^2, \dots), \, \dots, \, f_s = f(x_1^s, x_2^s, \dots)$$

are substituted in the polynomial $\Psi(f_1, f_2, \dots, f_s)$, *then* Ψ *turns into the generating function of the configurations of* [ϕ] *which are nonequivalent with respect to* $\mathfrak{H}$. *Such a polynomial* Ψ *must be the cycle index.*

The cycle index of $\mathfrak{H}$, we call it $\zeta(f_1, f_2, \dots, f_s)$, has the property imposed on $\Psi(f_1, f_2, \dots, f_s)$. It remains to prove the identity of the two polynomials ζ and Ψ; that is, that the coefficients in the expansions in powers of $f_1, f_2, \dots, f_s$ are the same.

We apply the assumption to the special collection of figures whose generating function is

$$f(x_1, x_2, \dots, x_n) = x_1 + x_2 + \cdots + x_n$$

(*n* balls of different colors; see Sec. 22). With the notation

$$(1.31) \qquad \begin{aligned} &f_1 = x_1 + x_2 + \cdots + x_n, \quad f_2 = x_1^2 + x_2^2 + \cdots x_n^2, \dots, \\ &f_s = x_1^s + x_2^s + \cdots + x_n^s \end{aligned}$$

the assumption implies that, in terms of the variables $x_1, x_2, \dots, x_n$, the two polynomials have the same coefficients, or, by (1.31)

$$(1.32) \qquad \Psi(f_1, f_2, \dots, f_s) - \zeta(f_1, f_2, \dots, f_s) = 0$$

identically in $x_1, x_2, \dots, x_n$. Choose $n \geqslant s$; it is well known[1] that for $s \leqslant n$ there can be no algebraic relationship between the first s power sums (1.31). Hence the left hand side of (1.32) must, as a polynomial in $f_1, f_2, \dots, f_s$, be identically zero.

26. Now we can look at the problem stated in Sec. 12 and solved in Sec. 19 from a different angle; namely, we can specify to what

---

[1]See, e.g., B. M. Bôcher, Einführung in die höhere Algebra (Leipzig, 1910, p. 263).

extent the solution depends on the structure of $\mathfrak{H}$. Two permutation groups of the same degree are labeled combinatorially equivalent if the solution of problem Sec. 12 is the same for any given collection of figures and any given content. (It goes without saying that the number of different colors is arbitrary, not restricted to three.) Specifically: two permutation groups, $\mathfrak{H}_1$ and $\mathfrak{H}_2$ of the same degree s are called combinatorially equivalent if the numbers of nonequivalent configurations derived from an arbitrary collection $[\phi]$ and with arbitrary content $(a_1, a_2, \dots)$ are the same for $\mathfrak{H}_1$ and $\mathfrak{H}_2$.

The main theorem, stated in Sec. 16 and proved in Sec. 19, combined with the proposition of Sec. 25 yields the following proposition: *Two permutation groups are combinatorially equivalent if and only if they have the same cycle index.*

Referring to the definition (1.5) of the cycle index we find further:[1] *two permutation groups are combinatorially equivalent if and only if there exists a unique correspondence between the permutations of the two groups such that corresponding permutations have the same type of cycle decomposition.*

It is of interest that two combinatorially equivalent permutation groups need not be identical.[2] They need not be isomorphic as abstract groups. Let p be an odd prime and m be an integer larger than 2 ($p = 3$, $m = 3$ furnishes the simplest example). It is known[3] that there exists a non-Abelian group of order p^m whose elements, except the identity, are of order p. Let $\mathfrak{H}_1$ denote the regular representation of this group as a permutation group and $\mathfrak{H}_2$ denote the regular representation of the Abelian group of order p^m and type $(p, \dots, p)$. $\mathfrak{H}_1$ and $\mathfrak{H}_2$ are permutation groups of order and degree p^m, and they are combinatorially equivalent: each of their permutations, except the identity, is decomposed into cycles the same way, in p^{m-1} cycles of length p, and the cycle index of both is

$$(f_1^{p^m} + (p^m - 1)f_p^{p^{m-1}})/p^m.$$

---

[1] If we restrict our attention to figures consisting of balls, as in Sec. 22, then the proposition is changed, with more restrictive necessary conditions and less restrictive conditions for sufficiency. The proof of this modified proposition is contained in our proof. Lunn and Senior, 1, p. 1053, state the proposition and provide a proof for sufficiency. Mr. Senior kindly communicated to me the second part (necessity) of the proof. The reasoning differs from the one presented in Sec. 25.

[2] Lunn and Senior 1, p. 1053.

[3] W. Burnside, *Theory of Groups of Finite Order*, 2nd ed. (Cambridge 1911), pp. 143-144.

27. We discuss some cases in which the cycle index of a group composed of several groups can be constructed from the cycle indices of the given groups in a transparent way.

Let $\mathfrak{G}$ and $\mathfrak{H}$ be two permutation groups with, respectively, orders g and h, degrees r and s, and cycle indices φ and ψ. Denote by $g_{i_1 i_2 \ldots i_r}$ (respectively $h_{j_1 j_2 \ldots j_s}$) the number of permutations in $\mathfrak{G}$ (respectively $\mathfrak{H}$) of type $[i_1, i_2, \ldots, i_r]$ (respectively $[j_1, j_2, \ldots, j_s]$). Then the cycle indices of $\mathfrak{G}$ and $\mathfrak{H}$ are

$$(1.33) \qquad \varphi = \frac{1}{g} \sum_{(i)} g_{i_1 i_2 \ldots i_r} \, f_1^{i_1} f_2^{i_2} \ldots f_r^{i_r} \,,$$

$$(1.34) \qquad \psi = \frac{1}{h} \sum_{(j)} h_{j_1 j_2 \ldots j_s} \, f_1^{j_1} f_2^{j_2} \ldots f_s^{j_s} \,.$$

Label the objects which $\mathfrak{G}$ and $\mathfrak{H}$ permute $x_1, x_2, \ldots, x_s$ and $y_1, y_2, \ldots, y_r$, respectively. Then the permutations of the two groups can be written as

$$(1.35) \qquad G = \begin{bmatrix} x_1, \ldots, x_\rho, \ldots, x_r \\ x_{1''}, \ldots, x_{\rho'}, \ldots, x_{r'} \end{bmatrix},$$

$$H = \begin{bmatrix} y_1, \ldots, y_\sigma, \ldots, y_s \\ y_{1''}, \ldots, y_{\sigma'}, \ldots, y_{s'} \end{bmatrix}.$$

We use $\mathfrak{G}$ and $\mathfrak{H}$ to construct two new permutation groups. The first is very simple and well-known, the second is more interesting.

The Direct Product. $\mathfrak{G} \times \mathfrak{H}$. Choose arbitrary permutations G and H, respectively, from $\mathfrak{G}$ and $\mathfrak{H}$. There are gh such pairs. Let the pair (G,H) correspond to the following permutation of the $r + s$ objects $x_1, x_2, \ldots, x_r, y_1, y_2, \ldots, y_s$:

$$\begin{bmatrix} x_1, x_2, \ldots, x_r & y_1, y_2, \ldots, y_s \\ x_{1''}, x_{2''}, \ldots, x_{r''} & y_{1''}, y_{2''}, \ldots, y_{s'} \end{bmatrix},$$

that is, the two permutations described by (1.35) are carried out simultaneously. It is obvious that the gh permutations of the $r + s$ objects form a permutation group, which we denote by $\mathfrak{G} \times \mathfrak{H}$ and call the direct product of $\mathfrak{G}$ and $\mathfrak{H}$. The product $\mathfrak{G} \times \mathfrak{H}$ is intransitive. Clearly (see Sec. 17), $\varphi\psi$ is the cycle index of $\mathfrak{G} \times \mathfrak{H}$.

The direct product $\mathfrak{G} \times \mathfrak{H} \times \mathfrak{K} \times \ldots$ of arbitrarily many permutation groups is defined similarly. With this definition of a direct product the degrees are added and the cycle indices multiplied.

The "corona" $\mathfrak{G}[\mathfrak{H}]$. Choose a permutation G from $\mathfrak{G}$ and any r permutations H_1, H_2, ..., H_ρ, ..., H_r from $\mathfrak{H}$; the H_ρ's need not be different. There are $g \cdot h^r$ different choices. As before, G is given by (1.35), and H_ρ is similarly defined by

$$(1.36) \qquad H_\rho = \begin{bmatrix} y_1, \; y_2, \; ..., \; y_s \\ y_{\rho_1}, \; y_{\rho_2}, \; ..., \; y_{\rho_s} \end{bmatrix} \qquad (\rho = 1, 2, ..., r).$$

We consider the following rs objects

$$z_{11}, z_{12}, ..., z_{1s} \; ,$$

$$z_{21}, z_{22}, ..., z_{2s} \; ,$$

$$(1.37) \qquad \cdot \; \cdot \; \cdot \; \cdot \; \cdot \; \cdot$$

$$z_{r1}, z_{r2}, ..., z_{rs} \; .$$

Let the $1 + r$ permutations G, H_1, H_2, ..., H_ρ, ..., H_r correspond to the following permutations of the rs objects:

$$\begin{bmatrix} z_{11}, \; ..., \; z_{1s}, \; ..., \; z_{\rho 1}, \; ..., \; z_{\rho s}, \; ..., \; z_{r1}, \; ..., \; z_{rs} \\ z_{1'1_1}, \; ..., \; z_{1'1_s}, \; ..., \; z_{\rho'\rho_1}, \; ..., \; z_{\rho'\rho_s}, \; ..., \; z_{r'r_1}, \; ..., \; z_{r'r_s} \end{bmatrix}.$$

That is, G is the "gross permutation",[1] it permutes the rows of the matrix (1.37); G indicates for every row to which row it is translated; while H_ρ defines the map of row ρ onto row ρ'. The gh^r permutations of the rs objects defined in this way form a group, which we denote by $\mathfrak{G}[\mathfrak{H}]$. We could call it corona of $\mathfrak{G}$ with respect to $\mathfrak{H}$.[2]

---

[1] I thank Mr. R. Remak for the fitting label.

[2] Geometric-kinematic example. A regular polyhedron has r faces. Each of the faces has e vertices and s ($\geqslant 1$) is an arbitrary multiple of e. The perpendicular line is mounted at the midpoint of each face and serves as the axis of a wheel. All r wheels are equal, each has s equidistant spokes, and each can be fixed in s positions: in each position a different spoke points to a certain vertex of the face in question. Suppose that the regular polyhedron allows g deck transformations. The deck transformations constitute a permutation group $\mathfrak{G}$ of order g and degree r for the r faces. The deck transformations of a wheel are the rotations by multiples of $2\pi/s$. The s spokes are subjected to a permutation group $\mathfrak{H}$; $\mathfrak{H}$ is cyclic of order and degree s. By combination of all deck transformations of the polyhedron and the wheels the rs spokes are permuted by a group of order gs^r and degree rs, specifically by the group of $\mathfrak{G}[\mathfrak{H}]$.

The permutations of $\mathfrak{G}[\mathfrak{H}]$ have a special effect on the rows of the matrix (1.37): if a permutation moves an element of one row into another row, then the permutation moves all the elements of the one row into the other row. The rows of (1.37) are imprimitive domains of $\mathfrak{G}[\mathfrak{H}]$. The permutations of $\mathfrak{G}[\mathfrak{H}]$ which leave the r imprimitive domains invariant (the gross permutation of which is the identity) form a subgroup; it has order h^r, it is the direct product $\mathfrak{H} \times \mathfrak{H} \times \mathfrak{H} \times \cdots \times \mathfrak{H}$ with r factors and is a normal subgroup of $\mathfrak{G}[\mathfrak{H}]$, with factor group,

$$\mathfrak{G}[\mathfrak{H}]/\mathfrak{H} \times \mathfrak{H} \times \cdots \times \mathfrak{H} = \mathfrak{G}.$$

Imagine the rs elements (1.37) as points in space and that a figure from $[\Phi]$ is attached to each point. This leads to the configuration

(1.38)
$$\begin{aligned} &\phi_{11}, \ \phi_{12}, \ ..., \ \phi_{1s}, \\ &\phi_{21}, \ \phi_{22}, \ ..., \ \phi_{2s}, \\ &\qquad \cdot \quad \cdot \quad \cdot \quad \cdot \quad \cdot \quad \cdot \\ &\phi_{r1}, \ \phi_{r2}, \ ..., \ \phi_{rs} \ . \end{aligned}$$

Each row is called a partial configuration. Two partial configurations

$$\phi_{\rho 1}, \ \phi_{\rho 2}, \ ..., \ \phi_{\rho s} \quad \text{and} \quad \phi_{\rho'1'}, \ \phi_{\rho'2'}, \ ..., \ \phi_{\rho's'}$$

are equivalent if there exists a permutation

$$H = \begin{bmatrix} 1, & 2 \ , & ..., & s \\ i_1, & i_2 \ , & ..., & i_s \end{bmatrix}$$

in $\mathfrak{H}$ such that

$$\phi_{\rho i_1} = \phi_{\rho'1'}, \ \phi_{\rho i_2} = \phi_{\rho'2'}, \ ..., \ \phi_{\rho i_s} = \phi_{\rho's'} \ .$$

(It does not matter whether ρ and ρ' are equal or different.) View all partial configurations which are equivalent to a given partial configuration as equal, as the same superfigure. The generating function of the various superfigures which can be extracted from the collection $[\Phi]$ can be derived from (1.34) by introducing the generating function of $[\Phi]$ according to the rules of Sec. 16.

The structure of the group $\mathfrak{G}[\mathfrak{H}]$ implies that two configurations of the form (1.38) are equivalent or nonequivalent with respect to $\mathfrak{G}[\mathfrak{H}]$ depending on whether the superfigures determined by the r rows form equivalent or nonequivalent configurations with respect to $\mathfrak{G}$. To determine the generating function of the nonequivalent configurations of rs figures ϕ modulo $\mathfrak{G}[\mathfrak{H}]$, we have to establish the generating function of the nonequivalent superfigures in (1.33) according to Sec. 16 (i.e., the function (1.34), where f stands for the generating function of $[\phi]$), and insert this function in (1.33), according to Sec. 16. Thus we form

$$(1.39) \qquad \psi_m = \frac{1}{h} \sum_{(j)} h_{j_1 j_2 \cdots j_s} \, f_m^{j_1} f_{2m}^{j_2} \cdots f_{sm}^{j_s}$$

$$(1.40) \qquad \varphi[\psi] = \frac{1}{g} \sum_{(i)} g_{i_1 i_2 \cdots i_r} \, \psi_1^{i_1} \cdots \psi_r^{i_r} \, ,$$

where f denotes the generating function of $[\phi]$.

The results of Sec. 25 imply: The cycle index of $\mathfrak{G}[\mathfrak{H}]$ is the expression $\varphi(\psi)$ given by (1.40), where ψ_1, ψ_2, ... denote polynomials of the independent variables f_1, f_2, ... as given in (1.39).

The following table provides some characteristics of the two permutation groups derived from the two groups $\mathfrak{G}$ and $\mathfrak{H}$.

Group	$\mathfrak{G}$	$\mathfrak{H}$	$\mathfrak{G} \times \mathfrak{H}$	$\mathfrak{G}[\mathfrak{H}]$
Degree	r	s	$r + s$	rs
Order	g	h	gh	gh^r
Cycle Index	φ	ψ	$\varphi\psi$	$\varphi[\psi]$

Chapter 2
GRAPHS

Definitions

28. In the next sections we describe in axiomatic-combinatorial terms what the chemists call structure and stereoformulas. To enhance the clarity of the exposition I provide more than the bare essentials. I begin by repeating some known definitions in graph theory. Some problems touched upon in the Introduction are going to be presented "officially" later on. I will adhere as much as possible to the terminology used by D. König in his elegant text.[1] I will highlight where substantial departure seemed to better serve the special purpose of this paper.

29. In the sequel, "graph" stands for "connected finite graph without loops." A graph is a system consisting of two kinds of elements, vertices and edges; the number of elements is finite; a relationship, called fundamental relationship is defined between a vertex and an edge. The fundamental relationship between the vertex P and the edge σ is given in terms borrowed from geometry: P is an endpoint of σ; σ starts in P, etc. The following two conditions are satisfied:

I.   Every edge is bounded by two vertices.

II.   By virture of the fundamental relationship, the elements of a graph, edges and vertices, form a connected system. In other words, any two vertices can be joined by a path consisting of a sequence of edges and vertices.

---

[1] König, 1. The reader is not expected to master the "theory of graphs" to understand this paper; it suffices to grasp the idea of graphs with the help of diagrams.

Condition I describes the fundamental relationship between an edge and exactly two points. The number of edges which relate to a given vertex is not restricted; it can be any natural number, zero included. A vertex which is not endpoint of an edge, is not related to any edge and thus not connected with any other element of the graph. Condition II implies that in this case the graph consists of this vertex. A graph which consists of a single vertex is called a single vertex graph.[1]

30. Consider an arbitrary graph with p vertices and s edges. The case $p = 0$, $s = 0$ (the null graph) is excluded from the discussion. If $s = 0$, that is, if there are no edges, different points cannot be connected. Condition II imposes $p = 1$, that is, if $s = 0$ we are necessarily dealing with a single point graph. For $s \geqslant 1$ condition II implies $p \geqslant 2$. Conditions I and II determine the following relationship between p and s:

$$(2.1) \qquad s - p + 1 = \mu,$$

where μ is non-negative; μ is called connectivity number of a graph.[2] A graph whose connectivity number equals 0 is called a tree; in other words, given the number p of vertices a tree is a graph with the smallest number of edges, namely $p - 1$.

A vertex which is endpoint of k edges is of degree (valence) k. A vertex of degree 1 is called an endpoint of the graph. Let p_k be the number of vertices of degree k. Except in the case of the single vertex graph for which $p_0 = 0$, we have

$$(2.2) \qquad p_0 + p_1 + p_2 + \cdots + p_k + \cdots = p\,,$$

and because of condition I,

$$(2.3) \qquad 0p_0 + 1p_1 + 2p_2 + \cdots + kp_k = 2s.$$

31. A tree consisting of more than one vertex and in which one endpoint plays a special role is called a planted tree.[3] The special endpoint is called "root" of the tree and the single edge which ends in the root is the stem of the planted tree. The vertices of a planted

[1]Introduction of the single vertex graph causes the most substantial departure from König's terminology.

[2]König, 1, p. 54.

[3]Not "rooted tree" (cf. König, 1, p. 76). I emphasize that not an arbitrary vertex but an endpoint is selected as root.

tree which are different from the root are called nodes.[1] In the figures of planted trees, nodes are denoted by circles, roots by arrows; see Figs. 1 - 3.

Trees without a special vertex, that is, without a root, are free trees or, simply, trees.

Let P be a vertex of a tree B and Q be the other endpoint of an edge (PQ) from P. The vertices P and Q together with other points of B which are not connected to P but to Q form with the connecting edges a planted tree with root P and stem PQ. This planted (sub)tree is a branch[2] of B at P or originating in P.

The following example may serve as a commentary to this definition: If B contains exactly p vertices then all the branches originating in P comprise together at least p vertices but exactly $p -$ 1 nodes. See Fig. 2 (α), (β), the branches originating in M.

32. The vertices of an arbitrary graph can be arbitrarily partitioned into species, subject to the obvious restriction that each vertex belongs to exactly one species. That is, two different species have no element in common. Imagine the vertices of one species as balls of the same color, or as atoms of the same element.

Planted trees have two types of vertices, roots and nodes, subject to two conditions: a root is the unique element of its species, it has valence one. In the general case there is no restriction. The partition into species can be arbitrary and is not tied to the valence. If the number of species is equal to the number of vertices, we are dealing with individually different vertices. At the other extreme is the graph in which all the vertices are interchangeable.

33. A point P of valence k forms together with the k edges a corona of edges;[3] P is its center. The k edges originating in P are numbered 1, 2, ..., k. I will list some examples of such numbering.

(a) Imagine the corona in the plane with the edges being straight line segments. The edges are numbered sequentially counterclockwise. Depending on the starting point, k different numbering schemes obtain. These k schemes are equivalent under the cyclic permutation groups $_k$ of order and degree k.

---

[1] The term "node" has a specific meaning and is used exclusively for vertices which are not roots of planted trees, in contrast to D. König's definition; König, 1, p. 1.

[2] A branch is always considered a planted tree. This definition deviates slightly from König's terminology, 1, p. 70.

[3] Not "star"; cf. König 1, p. 50. I emphasize that a corona contains a single point only. A corona is not a graph because its line segments are closed at one end only.

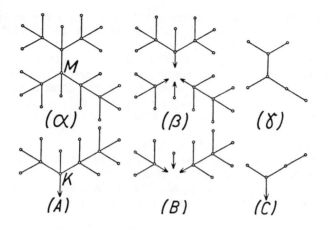

Figure 2

(b) Let $k = 4$. Suppose that the center of a corona coincides with the center of a regular tetrahedron and that the 4 edges connect the center with the vertices. Each edge is labelled by the same number as the corresponding endpoint. The tetrahedron vertices are numbered so that the tetrahedron carries the positive orientation. This means that a person, with the head in vertex 1 and the feet in vertex 2, facing edge 3 - 4 has 3 on the left and 4 on the right-hand side. In this way we can order the edges of the corona in 12 ways. It is easy to see that an even permutation of the vertices leaves the numbering "right-handed", an odd one produces a "left-handed" numbering.[1] Thus the 12 different numbering schemes are mapped into each other by the 12 permutations of the alternating group of degree 4, A_4.

(c) Again, we consider a corona of k edges, but this time we do it with respect to its topological qualities; that is, regardless of the space in which it is located. Then all $k!$ permutations of the symmetric group S_k are admissible.

Summing up the results: Depending on whether the corona of edges is treated as a configuration in two- or three-dimensional space or as a topological object, the admissible permutations form the "associated" groups Z_k, A_4, S_k, respectively.

The numbering of the edges and the proper permutation groups describe the problem completely and, therefore, it can be treated in purely combinatorial terms.

34. After this introduction we get to the main thrust of this paper. We are now in a position to determine completely and concisely,

---

[1] The 12 even permutations of the 4 vertices correspond to the rotations which are deck transformations of the regular tetrahedron.

without recourse to special cases, when two graphs are distinct and when they are not distinct. I will use the term "congruent" for not distinct, "noncongruent" for distinct.

Let G and G' be two graphs. The vertices of both are partitioned into species, the edges of all coronas are numbered,[1] for each k (k = 1, 2, ...) a group $\mathfrak{G}_k$ is given, associated with the coronas of degree k. The graphs G and G' are congruent if and only if there exists a one-to-one correspondence between G and G' by which

I.   each edge is mapped onto an edge;
II. each vertex is mapped onto a vertex of the same species;
III. the fundamental relationship is preserved;
IV. a permutation of the edge numbers is induced, and this permutation is in the associated group of each corona.

The last two conditions will be dealt with below in more detail. We call a mapping of G onto G' satisfying these four conditions a congruent mapping. Two elements which can be mapped onto each other may be called congruent; in this sense [(I), (II)] one may say:

Any two edges are congruent to each other. Vertices of the same species are congruent, vertices of different species noncongruent.

By condition III, that the fundamental relationship is preserved, we mean the following: if P and P' denote vertices, σ and σ' edges, P, σ belong to G and P', σ' to G', and if P is mapped to P', σ to σ', then P' is an endpoint of σ' if and only if P is an endpoint of σ.

Condition IV needs more explaining. Let P be the center of a k-edged corona of G and σ_1, σ_2, ..., σ_k be the numbered edges. Conditions I, II, III induce a mapping of these $k + 1$ elements of G onto the $k + 1$ elements P', σ_1', ..., σ_k' of G'. The numbering of the respective edges does not have to be the same. Let the correspondence be indicated by arrows:

$$P \to P'$$

$$\sigma_{i_1} \to \sigma_1', \quad \sigma_{i_2} \to \sigma_2', \quad ..., \quad \sigma_{i_k} \to \sigma_k'.$$

Then the induced permutation is understood to be

$$\begin{pmatrix} 1, & 2, & 3, & ..., & k \\ i_1, & i_2, & i_3, & ..., & i_k \end{pmatrix}.$$

---

[1] With complete numbering of the coronas each edge is assigned two unrelated numbers.

Condition IV means that this permutation belongs to the group $\mathfrak{C}_k$ which is associated with the coronas of degree k.

Re-examining the four conditions, one notices that the congruence of graphs as defined here establishes a reflexive, symmetric, and transitive relation between graphs. Observe that (IV) does not involve an arbitrary set but a group of permutations.

35. As far as I can judge, the definition of congruence and non-congruence (nondisparity, disparity) of two graphs is essential to fix the meaning of chemical formulas, especially "stereoformulas". I restrict attention to the meaning of the numbers which are calculated in the present chapter with respect to this definition. I will introduce chemical terminology only in the next chapter, and proceed here with geometric-combinatorial considerations, including comparisons with the planar realization of graphs which is uninteresting from the chemical point of view.

The given definition of congruence of two graphs contains many special cases. Specialization can go in three directions: one considers special graphs; the vertices are subdivided in a special way into species; special permutation groups associated with the coronas are given. The following terminology serves the threefold specialization.

A C-graph is a graph in which no vertex is endpoint of five or more edges. In the notation of Sec. 30, a C-graph is characterized by

$$(2.4) \qquad p_5 = p_6 = p_7 = \cdots = 0.$$

A C-H graph contains only vertices of degree 1 (endpoints) and degree 4, that is,

$$(2.5) \qquad p_0 = 0, \, p_2 = p_3 = 0, \, p_5 = p_6 = p_7 = \cdots = 0.$$

Unless stated otherwise, we consider all the vertices of a free tree to be of the same species (Sec. 45 in this chapter is the only exception).

In planted trees we distinguish between two species of points, the nodes and the root; that is, all nodes are considered to be of the same species unless stated otherwise (as in Sec. 45).

We consider only the groups $\mathcal{Z}_k$, $\mathcal{A}_4$, and $\mathcal{S}_k$. If $\mathcal{S}_k$ is the group associated with coronas of degree k $(k = 1, 2, 3, ...)$, then two graphs which are congruent according to Sec. 34 are topologically nondistinct; if they are noncongruent then they are topologically distinct.

If the cyclic group $\mathcal{Z}_k$ is associated with the coronas of order k $(k = 1, 2, ...)$ then two graphs which are congruent according to the definition of Sec. 34 are called "planar nondistinct", while if they are noncongruent in this sense they are called "planar distinct".[1]

---

[1]Whether it is possible to sketch a graph on a piece of paper in such a way that no edges intersect which have no point in common is irrelevant. A graph in the plane has nothing to do with a graph of genus 0. (Cf. König 1, p. 198).

Congruence of C-H graphs can be defined by associating the alternating group A_4 with the coronas of degree 4. (This is true for C-H graphs only.) For coronas of degree 1 condition (IV) adds nothing to (I), (II), (III); the permutation group of degree 1 consists of the identity. Two C-H graphs are spatially nondistinct if they are congruent under A_4; if they are noncongruent they are spatially distinct.

The difference between free and planted trees has to do with condition (II) of Sec. 34. For free trees (II) can be restated: any two vertices are congruent; for planted trees: nodes are congruent among each other, roots are congruent among each other; a node is never congruent to a root. The partition into planar, spatial, topological graphs is tied to condition (IV); for topological graphs it does not add anything.

36.   In light of the above expositions the definitions given in the Introduction of the numbers τ_n and T_n have a purely combinatorial meaning.   The other quantities introduced, κ_n, ρ_n, σ_n, Q_n, R_n, S_n, are tied to C-H trees, and thus, they have to be discussed further.

(a) Let n denote the number of vertices of degree 4 of a C-H graph; in the notation of Sec. 30, this means $p_4 = n$. Equation (2.5) combined with (2.1), (2.2), and (2.3) leads to

(2.6) $p_1 = 2n + 2 - 2\mu.$

Specifically, a C-H graph with n vertices of degree 4 is a C-H tree if and only if the number of endpoints equals $2n + 2$ and the total number of points equals $3n + 2$. A planted C-H tree with n vertices of degree 4 has $2n + 1$ endpoints which are different from the root, hence a total of $3n + 1$ nodes.

(b) Certain vertices of degree 4 of a C-H tree are asymmetric, namely those from which 4 topologically different branches originate (e.g., the point M in Fig. 2 (α) is asymmetric). The definition of an asymmetric vertex of a planted C-H tree is the same except for the qualifier: The branch originating in the point P of degree 4 which contains the root of the tree is by definition different from any other branch from P: it carries the root which is noncongruent to the vertices of all 3 other branches. (The point K of the planted tree in Fig. 2(a) is asymmetric; if it were a free tree, not a planted tree, then K would be symmetric.)

The numbers σ_n, ρ_n, κ_n refer to free C-H trees, the numbers S_n, R_n, Q_n to planted C-H trees with n vertices of degree 4:

σ_n (resp. S_n) the number of all spatially different trees;

ρ_n (resp. R_n) the number of all topologically different trees;

κ_n (resp. Q_n) the number of all topologically different trees which have no asymmetric vertices.

(See Sec. 55 for the equivalence of these combinatorial definitions and the chemical ones given in the Introduction.) The following relations will help to elucidate the definitions.

The transitions from ρ_n to κ_n as well as from R_n to Q_n are transitions from a set to a subset. Hence,

$$(2.7) \qquad \rho_n \geqslant \kappa_n, \quad R_n \geqslant Q_n.$$

The transition from ρ_n to σ_n as well as from R_n to S_n consists in replacing the permutation groups associated with the coronas of degree 4, replacing $\mathcal{S}_4$ with A_4. The condition that the permutation induced by the mapping (see Sec. 34, condition (IV)) belong to A_4 is more restrictive than that it belong to $\mathcal{S}_4$. By restricting the group to a subgroup we cannot end up with fewer nonequivalent configurations. Hence,

$$(2.8) \qquad \rho_n \leqslant \sigma_n, \quad R_n \leqslant S_n.$$

37. Cayley has pointed out that there exists a one-to-one relationship between topological C-H graphs and C-graphs which can be explained as follows: Delete all endpoints and the corresponding edge, except the root and the stem in the case of a planted tree. What is left is the associated C-tree, specifically, the associated tree of a planted tree has the same root and stem. (In Fig. 2, move from (α) to (γ) and from (a) to (c).) The C-H graph can be reconstructed from the C-graph by adding to each vertex of the C-graph except to the possible root, so many edges that it turns into a point of degree 4. Each of the new edges is capped by an endpoint. (In Fig. 2 return from (γ) to (α), from (c) to (a).)[1]

Pursuing this correspondence between a C-H graph and the corresponding C-graph we find a new interpretation of the numbers ρ_n and R_n: ρ_n is the number of topologically different free C-trees with n vertices, R_n is the number of planted C-trees with n nodes. In other words:

---

[1] If the C-H graph contains n vertices of degree 4, the associated C-graph consists of n points, respectively $n + 1$ points in the case of a planted tree. In the first case, $n = 0$ is not admissible; in the second (planted trees) it is. With this agreement the one-to-one correspondence holds throughout, and the special case of the null graph is included. In the reconstitution of the C-H graph from the C-graph, 4 edges are added only in the case of the single-vertex graph.

ρ_n is the number of topologically different free trees with n vertices each of which is an endpoint of at most 4 edges;

R_n is the number of topologically different planted trees with n nodes, each of which is an endpoint of at most 4 edges.

Compare this with the definitions of the numbers τ_n and T_n in the Introduction (Sec. 3). The transition from τ_n to ρ_n as well as from T_n to R_n is the transition from a set to a subset, whence

(2.9) $\tau_n \geqslant \rho_n, \quad T_n \geqslant R_n.$

Planted Trees

38. Henceforth we assume that the group $\mathfrak{C}_k$ associated with the coronas of degree k is transitive ($k = 1, 2, ...$). This assumption is satisfied for the groups $\mathfrak{S}_k$, $\mathfrak{Z}_k$, $\mathfrak{A}_4$ which characterize the topological, the planar, and the spatial arrangement.

The stem of a planted tree S is bounded by two vertices, one is the root, the other the principal node, K for short (see Fig. 2(a)). Let K be a vertex of degree k. One of the k branches originating in K (see Sec. 31) contains the stem and the root of S (in everyday language this would not be a branch). We call the other $k - 1$ branches principal branches of the planted tree S. We number the principal branches: Each principal branch contains an edge originating in K which, as an element of the corona of edges from K, carries a number; the same number is assigned to the corresponding principal branch.

I claim that one can find a planted tree S' which is congruent to S and whose principal branches are numbered 1, 2, ..., $k - 1$. This is a consequence of the transitivity of the group $\mathfrak{C}_k$. This group contains a permutation which maps the number assigned in the corona of K to the stem of S into k. Subject the corona of K to this permutation and leave everything else unchanged. The resulting planted tree S' has the desired properties (cf. Sec. 34, in particular (IV)).

39. Consider planted trees whose $k - 1$ principal branches carry the numbers 1, 2, ..., $k - 1$. Denote the principal branches of S by $\phi_1, \phi_2, ..., \phi_{k-1}$, with the indices corresponding to the numbering. Assign to S the configuration of its principal branches,

$$(\phi_1, \phi_2, ..., \phi_{k-1}).$$

The name "configuration" has here the same connotation as in Sec. 11.[1] Congruent principal branches are considered equivalent, non-

---

[1] If we want to retain the spatial image of Sec. 11, we may assign $k-1$ points in space to the $k-1$ different endpoints of the $k-1$ principal branches near, but different from, K, i.e. to their roots.

congruent principal branches of different figures. Since each principal branch is a planted tree (Sec. 31), the collection of figures contains all (pairwise noncongruent) planted trees.

Two planted trees whose principal branches form the same configuration are certainly congruent in the sense of Sec. 34.[1] Is it possible that different configurations of the principal branches belong to congruent planted trees?

The letters K', k', ϕ'_1, ϕ'_2, ..., ϕ'_{k-1} mean for S' what K, k, ϕ_1, ϕ_2, ..., ϕ_{k-1} mean for S. If S and S' are congruent to each other (see conditions (I), (II), (III), (IV) in Sec. 34) then the root of S has to correspond to the root of S', the stem of S to the stem of S', and K to K', hence $k = k'$. Since in the numbering of the coronas around K and K', respectively, the stems of S and S' carry the number k, the induced permutation must be of the form

$$(2.10) \qquad \begin{bmatrix} 1 & 2 & ... & k-1 & k \\ i_1 & i_2 & ... & i_{k-1} & k \end{bmatrix},$$

and it must belong to the group $\mathfrak{G}_k$ (Sec. 34 (IV)). Finally, because congruent principal branches are, as figures, identical, we must have

$$(2.11) \qquad \phi'_1 = \phi_{i_1}, \quad \phi'_2 = \phi_{i_2}, \quad ..., \quad \phi'_{k-1} = \phi_{i_{k-1}}.$$

The permutation (2.10) belongs to $\mathfrak{G}_k$, specifically it belongs to the subgroup of $\mathfrak{G}_k$ which leaves k invariant. This subgroup, which is a permutation group of degree $k - 1$ (it permutes 1, 2, ..., $k-1$) is called the associated subgroup. The equations (2.11) can now be stated in terms of Sec. 11: the configurations $(\phi_1, \phi_2, ..., \phi_{k-1})$ and $(\phi'_1, \phi'_2, ..., \phi'_{k-1})$ are equivalent with respect to the associated subgroup.

Reading the deductions in reverse order, we conclude: two planted trees are congruent to each other if and only if they have the same number of principal branches and the configurations of the principal branches are equivalent with respect to the associated subgroup. Depending on the associated groups $\mathfrak{S}_k$, $\mathfrak{A}_4$, $\mathfrak{Z}_k$, the associated subgroups are $\mathfrak{S}_{k-1}$, $\mathfrak{A}_3 = \mathfrak{Z}_3$, and $\mathfrak{Z}_{k-1}$, respectively. These cases are of import below.

40. We examine the planted C-H trees with n vertices of degree 4: the number of topologically different trees is R_n, of spatially different trees S_n, and of two-dimensionally different trees P_n. The

---

[1]For a proof of this statement, one has to consider two planted trees, S and S', whose principal branches, numbered in the same way, are congruent with each other. The $k - 1$ congruent mappings of the principal branches are used to construct a congruent mapping from S to S'. The details are left to the reader.

principal node of a planted C-H tree is either an endpoint or of degree 4.

If the principal node K is an endpoint, the planted tree consists of K, the root and the stem, which connects these two points. There are no vertices of degree 4, there are no principal branches. There are no two noncongruent planted trees of this type, whether we deal with topological, spatial, or planar congruence. Hence

(2.12) $R_0 = S_0 = P_0 = 1.$

If the principal node of a planted C-H tree S has valence 4, then the tree has 3 principal branches, and together they contain exactly one vertex of degree 4 less than S does. Hence, there are, for $n \geqslant 1$, exactly the same number of noncongruent planted C-H trees with n nodes of degree 4 as there are nonequivalent configurations, with respect to the associated subgroup, of three planted C-H trees containing a total of $n - 1$ four-edged nodes. (Cf. Fig. 2(a) and (b); depending on the topological, spatial, or planar interpretation of the congruence, the three given mutually different principal branches give rise to 1, 2, or 6 different configurations.)

According to the nature of the congruence, topological, spatial, or planar, the generating functions of the planted C-H trees are given by

$$r(x) = \sum_0^\infty R_n x^n, \quad s(x) = \sum_0^\infty S_n x^n, \quad p(x) = \sum_0^\infty P_n x^n;$$

the associated subgroups are

$$\mathcal{S}_3, \quad \mathcal{A}_3, \quad \mathcal{E}_3 ;$$

the cycle indices of these subgroups are

$$(f_1^3 + 3f_1 f_2 + 2f_3)/6, \quad (f_1^3 + 2f_3)/3, \quad f_1^3 .$$

Making use of the relationship discussed above, "the number of non-congruent planted trees equals the number of nonequivalent configurations of three planted trees", of the generating function and the main theorem of Chapter 1 (Sec. 16) and taking the special case $n = 0$ into account, we establish for each of the three situations an equation:

(2.13) $r(x) = R_0 + x(r(x)^3 + 3r(x)\, r(x^2) + 2r(x^3))/6,$

(2.14) $s(x) = S_0 + x(s(x)^3 + 2s(x^3))/3,$

(2.15) $p(x) = P_0 + xp(x)^3.$

41. Together with equation (2.12), equations (2.13) and (2.14) are identical with (4) and (7). When P_0 from (2.12) is substituted in (2.15) a trinomial equation of third degree for $p(x)$ obtains. The solution can be given in terms of a power series[1]

$$p(x) = 1 + \sum_{n=1}^{\infty} \binom{3n}{n-1} x^n/n.$$

Hence, we have

$$(2.16) \qquad P_n = \frac{1}{n} \binom{3n}{n-1}.$$

Noting that the transition from S_n to P_n corresponds to the transition from $\mathbf{A}_3$ to $\mathbf{E}_3$, from the group to a subgroup, we find, as in (2.8),

$$(2.17) \qquad S_n \leqslant P_n.$$

42. Again, we are dealing with planted C-H trees with n vertices of degree 4. We restrict our attention to topologically different ones with exactly α asymmetric points; let $R_{n\alpha}$ denote their number. Obviously, the relation

$$(2.18) \qquad R_{n0} + R_{n1} + R_{n2} + \cdots = R_n$$

holds, and by an earlier definition (Sec. 36)

$$(2.19) \qquad R_{n0} = Q_n.$$

We write

$$(2.20) \qquad \sum_{n=0}^{\infty} \sum_{\alpha=0}^{\infty} R_{n\alpha} x^n y^\alpha = \sum_{n=0}^{\infty} x^n (R_{n0} + R_{n1} y + R_{n2} y^2 + \cdots)$$
$$= \phi(x,y).$$

Clearly,

$$(2.21) \qquad R_0 = R_{00} = 1.$$

Therefore, we assume $n \geqslant 1$ and we consider here planted trees among the total of R_n trees. Each such planted tree is represented by the configuration of its three principal branches; specifically, since $\mathbf{S}_3$ is the associated subgroup, we have to know only which planted trees appear as principal branches. The numbering does not play any role; that is, only the combination of the three principal branches matters.

---

[1] Cf. e.g., Pólya and Szegö, *Problems and Theorems in Analysis*, Vol. 1, Problem III, 211, pp. 146, 348.

A planted tree counted in the number $R_{n\alpha}$ belongs to one of the following two types:

(1) The principal node K is not an asymmetric point; in this case the three principal branches are not all different and they have a total of α asymmetric vertices.
(2) The principal node K is asymmetric; in this case the three principal branches are all different and they have a total of $\alpha - 1$ asymmetric vertices.

In both cases the three principal branches contain $n - 1$ nodes of degree 4. Taking the special case $n = 0$ into account and using the results of Sec. 23 [$(2\ \mathcal{S}_3 - A_3)$ applies to case (1) and $(A_3 - \mathcal{S}_3)$ to case 2)], we find

(2.22)
$$\phi(x,y) = 1 + x\phi(x,y)\phi(x^2,y^2)$$
$$+ xy(\phi(x,y)^3 - 3\phi(x,y)\phi(x^2,y^2) + 2\phi(x^3,y^3))/6.$$

Equations (2.18), (2.20), (3) imply

(2.23) $\phi(x,1) = r(x)$,

and (2.19), (2.20), (6) imply

(2.24) $\phi(x,0) = q(x)$.

In fact, the functional equation (2.22) reduces for $y = 1$ to (4), and for $y = 0$ to (8).

43. Now we turn to arbitrary planted trees with a total of n nodes: T_n denotes the number of topologically different trees, P_n the number of two-dimensionally different trees. In addition to the generating function (2), we consider

$$\overline{p}(x) = \overline{P}_1 x + \overline{P}_2 x^2 + \overline{P}_3 x^3 + \cdots .$$

It is easy to see [same figure as for (2.12)] that

(2.25) $T_1 = \overline{P}_1 = 1$.

For $n \geqslant 2$ the planted tree has principal branches; let $k - 1$ be their number, as in Sec. 38. These $k - 1$ principal branches contain a total of $n - 1$ nodes, the subgroup associated with their configuration is $\mathcal{S}_{k-1}$ or $\mathcal{E}_{k-1}$, depending on whether T_n or P_n is involved. The number of configurations of principal branches which are non-equivalent with respect to $\mathcal{S}_{k-1}$ is, according to Sec. 16, the coefficient of x^{n-1} in the series which obtains upon substitution of $t(x)$ in the cycle index of $\mathcal{S}_{k-1}$. Compare formula (1.12) and, for

special cases, Sec. 21. The number of configurations which are nonequivalent with respect to $\mathfrak{E}_{k-1}$ is the coefficient of x^{n-1} in the
series $t(x)^{k-1}$. Introducing $k = 2, 3, 4, \dots$ and taking the special case
of $n = 1$ and (2.25) into account, one finds

(2.26)
$$t(x) = x + xt(x) + x(t(x)^2 + t(x^2))/2$$
$$+ x(t(x)^3 + 3t(x)\, t(x^2) + 2t(x^3))/6 + \cdots ,$$

(2.27) $$\bar{p}(x) = x + x\bar{p}(x) + x\bar{p}(x)^2 + x\bar{p}(x)^3 + \cdots .$$

The left-hand sides count the noncongruent planted trees, the right-
hand sides the principal branch configurations which are nonequivalent with respect to the associated subgroups. They are configurations of planted trees of the same type and, according to Sec.
39, there are exactly as many as there are noncongruent planted
trees.

Relation (2.26) is the one mentioned in the Introduction (1″).
Specialize formula (1.11) in two ways: firstly, set $f(x,y,z) = t(x)$ and
correspondingly

$$a_{000} = 0, \quad a_{k00} = T_k, \quad a_{k\ell m} = 0 \quad \text{for} \quad \ell + m > 0.$$

Secondly, let $u = 1$; then comparison of the first, second, and third
line of (1.11) yields the other two equations (1) and (1′).

Equation (2.27) is equivalent to an equation of second degree for
$\bar{p}(x)$,

$$\bar{p}(x) - \bar{p}(x)^2 = x,$$

which is satisfied by the series

$$\bar{p}(x) = \sum_{n=1}^{\infty} \binom{2n-2}{n-1} \frac{x^n}{n} .$$

Therefore

(2.28) $$\bar{P}_n = \frac{1}{n} \binom{2n-2}{n-1} .$$

The transition from T_n to $\bar{P}_n$ corresponds to the transition from $\mathfrak{S}_{k-1}$
to $\mathfrak{E}_{k-1}$, that is, from group to subgroup; accordingly, we have
(similar to (2.8) and (2.17))

(2.29) $$T_n \leqslant \bar{P}_n .$$

The transition of $\bar{P}_{3n+1}$ to P_n is a transition from a set to a subset;
hence [it is easy to see that there is no contradiction to (2.16) and
(2.28)]

$$\bar{P}_{3n+1} \geqslant P_n .$$

44. The number of topologically different planted C-trees with n nodes is R_n ($n \geqslant 1$), as we have noted in Sec. 37. It is easy to see that

$$R_1 = 1$$

[as in (2.25)]. For $n \geqslant 2$ the planted C-tree has 1, 2, or 3 principal branches. The reasoning which yielded (2.26) leads for the series

$$R_1 x + R_2 x^2 + R_3 x^3 + \cdots = r(x) - 1 = g(x)$$

to the functional equation

(2.30)
$$g(x) = x + xg(x) + x(g(x)^2 + g(x^2))/2$$
$$+ x(g(x)^3 + 3g(x)g(x^2) + 2g(x^3))/6.$$

The right-hand side of (2.30) contains only four terms; they correspond to the four possible cases of a rooted C-tree: there are 0, 1, 2, or 3 principal branches. (The right-hand side of (2.26) contains infinitely many terms.) Substituting $g(x) = r(x) - 1$ in (2.30), we obtain equation (4) as expected.

A slight generalization of the reasoning shows that all equations, which are derived from (2.26) or (2.27) through elimination of some of the terms on the right-hand side which are not identical to x, reduce to generating functions of easily characterizable types of planted trees.[1]

45. Now we examine the structure of (1') and (1'') of equation (1) from a different angle. We consider free trees with n mutually different vertices; α_n is the number of different trees. Cayley was the first to calculate α_n; others followed.[2] Here we present a new (as far as I know) calculation of α_n.

Each tree of the type described here gives rise to a planted tree by addition of an edge to an arbitrarily selected vertex. The added endpoint is declared the root. Since the n vertices are different we

---

[1]For example, by retaining only one term which differs from x we derive from (2.27) $y(x) = x + xy(x)^2$. The solution of this equation represents the different planar planted trees with n nodes which have only vertices of degrees 1 and 3. Cayley already established this result, 2. Cf. F. Levi, Christiaan Huygens, 2 (1922), pp. 307-314, No. 5, further A. Errera, Mémoires de l'Académie royale de Belgique 11, 1931, pp. 1-26, No. 15-16. Both contain (2.28) in different form and investigations of planar free trees.

[2]Cayley, 8, D. Dziobek, Sitzungsber. d. Berliner Math. Ges. 16(1917), pp. 64-67, H. Prüfer, Archiv. d. Math. u. Physik (3) 27(1918), pp. 142-144.

can construct n different planted trees. Let A_n be the number of topologically different planted trees with n mutually different nodes. Then the above considerations imply

(2.31) $A_n = n\alpha_n$.

Obviously,

(2.32) $A_1 = 1$.

Examine the planted trees with $n + 1$ distinct nodes which satisfy the following two conditions:

(1) The "red point" plays the role of K, the principal node.
(2) There are three principal branches originating in K.

(I use "red point" instead of "a certain point" to emphasize its special role; the number of principal branches is set at three only to fix the ideas.) If the three principal branches at K have i, j, and k, nodes, respectively, then

(2.33) $i + j + k = n$.

The number of different assignments of n distinguishable elements into three classes of i, j, and k elements is given by

$$\frac{n!}{i! \, j! \, k!} .$$

Once the classification is fixed, the first principal branch can be chosen in A_i, the second in A_j, and the third in A_k different ways. Thus the number of configurations of principal branches is

$$\sum_{i+j+k=n} \frac{n!}{i! \, j! \, k!} A_i A_j A_k \, ,$$

with the summation over all triples i, j, k which satisfy (2.33). Of these configurations $3!$ are equivalent with respect to $\mathcal{S}_3$,[1] hence there are

(2.34) $$\frac{1}{3!} \sum_{i+j+k=n} \frac{n!}{i! \, j! \, k!} A_i A_j A_k$$

configurations of principal branches which are nonequivalent with respect to $\mathcal{S}_3$; that is, the number is equal to the number of the planted trees counted by A_{n+1}, which satisfy the conditions (1) and (2). By removing (1), that is, by successively choosing every vertex

---

[1] Since the nodes are mutually different, permutation of the principal branches cannot leave the configurations invariant (except the identity).

as "red point", we get $(n + 1)$ times the number (2.34), and by removing (2) in addition to (1), that is, by considering arbitrarily many principal branches, we find all planted trees counted by A_{n+1}. That is,

(2.35)

$$A_{n+1} = (n + 1)A_n + \frac{n + 1}{2!} \sum_{i+j=n} \frac{n!}{i! \, j!} A_i A_j$$

$$+ \frac{n + 1}{3!} \sum_{i+j+k=n} \left[\frac{n!}{i! \, j! \, k!} \right] A_i A_j A_k + \cdots .$$

The terms on the right-hand side correspond to the different possible cases of 1, 2, 3, ... principal branches. Introducing the generating function

$$f(x) = \frac{A_1 x}{1!} + \frac{A_2 x^2}{2!} + \frac{A_3 x^3}{3!} + \cdots ,$$

and taking the case of no principal branch into account (cf. (2.32)), we obtain

(2.36) $$f(x) = x + xf(x) + x \frac{f(x)^2}{2!} + x \frac{f(x)^3}{3!} + \cdots .$$

We see this by comparing the coefficients of x^{n+1} on the left- and right-hand sides of (2.36), which leads (for $n + 1 \geq 2$) to the equation (2.35) divided by $(n + 1)!$. Therefore, (2.36) becomes[1]

(2.37) $$f(x) = xe^{f(x)}.$$

The solution of this functional equation is given by the series

$$f(x) = \frac{x}{1!} + \frac{(2x)^2}{2!2} + \cdots + \frac{(nx)^n}{n!n} + \cdots .$$

Thus, we find

(2.38) $$A_n = n^{n-1},$$

and because of (2.31)

$$\alpha_n = n^{n-2} ;$$

the second is Cayley's surprisingly simple result.

   A relationship between the numbers A_n and T_n corresponds to the similarity of the two equations (2.36) and (1'') [of (2.37) and (1')]. Choose a topological planted tree counted by T_n with n nodes of the same species. Then label these n nodes individually. The resulting

---

[1]Cf. e.g., Pólya and Szegö, *Problems and Theorems in Analysis*, Vol. 1, Problem III, 209, pp. 146, 348.

tree is counted by A_n. By permuting the labels we obtain not necessarily all topologically different planted trees counted by A_n.[1]
Hence,

$$(2.39) \qquad n! \, T_n \geqslant A_n = n^{n-1}$$

and similarly,

$$(2.40) \qquad n! \, T_n \geqslant \alpha_n = n^{n-2}.$$

Trees

46. Let B be a tree with n vertices. The vertices of B are partitioned into two classes, in ordinary vertices and exceptional vertices.
A point P of B is called an ordinary point if a branch with more
than $n/2$ nodes originates in P; a vertex is exceptional if it is not
ordinary. C. Jordan has formulated the following proposition:[2]

A tree with n vertices has either one or two exceptional points.

*If there is only one exceptional point M then no branch with at least
n/2 nodes originates in M.*

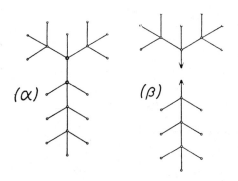

Figure 3

[1]The details of a similar conclusion are hinted at in Sec. 54a.

[2]Jordan 1; cf. König 1, pp. 70-75.

If there are two exceptional points M_1 and M_2, then the number n is even and both vertices, M_1 and M_2, carry branches of n/2 nodes each, and M_1 and M_2 are the endpoints of a certain edge.

Trees with a single exceptional point are centric and the exceptional point is the center of the tree. Trees with two exceptional vertices, called bicenters, are bicentric and the connecting edge is the axis of the tree.[1] (Figure 2(α) shows a centric, Fig. 3(α) a bicentric tree.) The simplest centric tree is a single vertex graph; it consists of the center. The simplest bicentric tree has two points; it consists of the two bicenters and the connecting axis.

47. In a congruent mapping of the tree B onto the tree B' the point P' of B' corresponds to the point P of B. Then each branch from P' corresponds to a branch originating in P, specifically a branch with the same number of nodes (Sec. 34 (I), (II), (III); (IV) does not play a role yet). It follows that in congruent mappings, exceptional points correspond to each other. A centric tree can be congruent to a centric tree only, and a bicentric tree can be congruent to a bicentric tree only.

For this reason we want to determine the number of centric and bicentric trees separately. Numbers of centric trees will be distinguished from those of bicentric trees by primes and double primes, respectively. Of the ρ_n, σ_n, τ_n, respectively, free trees ρ_n', σ_n', τ_n' are centric, and ρ_n'', σ_n'', τ_n'' are bicentric, such that

$$(2.41) \qquad \rho_n = \rho_n' + \rho_n'', \quad \sigma_n = \sigma_n' + \sigma_n'', \quad \tau_n = \tau_n' + \tau_n''.$$

We call $\rho_{n\alpha}$ the number of topologically distinct free C-H trees with n vertices of degree 4, which have exactly α asymmetric points. Suppose $\rho_{n\alpha}'$ are centric and $\rho_{n\alpha}''$ are bicentric, such that

$$(2.42) \qquad \rho_{n\alpha} = \rho_{n\alpha}' + \rho_{n\alpha}''$$

$$(2.43) \qquad \begin{aligned} \kappa_n &= \rho_{n0}' + \rho_{n0}'' \\ \rho_n &= \rho_{n0} + \rho_{n1} + \rho_{n2} + \cdots. \end{aligned}$$

48. Let B denote a bicentric tree with n points, M_1 and M_2 its bicenters, ϕ_1 and ϕ_2 the branches with n/2 nodes originating in M_1 and M_2, respectively; M_1', M_2', ϕ_1', ϕ_2' have the corresponding meaning for the tree B' (see Figure 3). The two trees B and B' are congruent if and only if one of the following two (not necessarily exclusive) situations occurs: Either ϕ_1 is congruent to ϕ_1' and ϕ_2 to ϕ_2', or ϕ_1 is congruent to ϕ_2' and ϕ_2 to ϕ_1'. Consequently, the number of bicentric free trees with n points is equal to the number of (not ordered) pairs of planted trees with n/2 nodes.

---

[1] König 1, p. 73, uses the more precise (here not necessary) labels of masscenter, massbicenter, massaxis.

The following are special cases:

$$(2.44) \qquad \tau_n'' = \frac{1}{2} T_{n/2}(T_{n/2} + 1),$$

$$(2.45) \qquad \sigma_n'' = \frac{1}{2} S_{n/2}(S_{n/2} + 1),$$

$$(2.46) \qquad \rho_n'' = \frac{1}{2} R_{n/2}(R_{n/2} + 1),$$

$$(2.47) \qquad \rho_{n0}'' + \rho_{n1}''y + \rho_{n2}''y^2 + \cdots$$

$$= \frac{1}{2} [(R_{\frac{n}{2},0} + R_{\frac{n}{2},1}y + R_{\frac{n}{2},2}y^2 + \cdots)^2$$

$$+ R_{\frac{n}{2},0} + R_{\frac{n}{2},1}y^2 + R_{\frac{n}{2},2}y^4 + \cdots],$$

$$(2.48) \qquad \rho_{n0}'' = \frac{1}{2} Q_{n/2}(Q_{n/2} + 1).$$

In the last four formulas, n stands for the number of vertices of degree 4; the number of vertices of this tree is $3n + 2$. For odd n, both sides of the five formulas may be assigned the value 0. With this agreement they hold for all n, $n = 1, 2, 3, \ldots$. Relation (2.47) is derived by means of the main theorem of Chapter 1, applied to the special case of $\mathfrak{S}_2$.

49. Let B be a centric tree, whose center M is a vertex of degree k, and ϕ_1, ϕ_2, ..., ϕ_k are the branches originating in M. Each of the branches carries the number the edge is assigned to in the corona of M. We consider the configuration

$$(\phi_1, \phi_2, \phi_3, \ldots, \phi_k).$$

Similar arguments as in Sec. 39 lead to the following result: Two centric trees are congruent if and only if the same number of branches originates from their centers and the configurations of these branches are equivalent with respect to the group associated with the corona of the center.

Accordingly, counting the noncongruent free trees of a certain type is reduced to counting the nonequivalent configurations of the planted trees of the corresponding type, in particular the determination of ρ_n' is reduced to the computation of R_n, σ_n' to that of S_n, τ_n' to that of T_n, as we will see in more detail below.

50. I will use the following abbreviations in the sequel:

$$f(x) = a_0 + a_1 x + a_2 x^2 + \cdots$$

denotes an arbitrary power series. The sum of the first $n + 1$ terms is written as

$$a_0 + a_n x + \cdots + a_n x^n = f(x)$$

and the n-th coefficient of $f(x)$ as

$$a_n = \text{Coeff}_n \, f(x).$$

Let m be the maximum number of nodes of a branch which originates in the center of a centric tree with n vertices. In other words, m is the integer which satisfies the double inequality

(2.49) $$\frac{n}{2} - 1 \leqslant m < \frac{n}{2}.$$

51. We consider free centric C-H trees with n vertices of degree 4, that is, with a total of $3n + 2$ points [Sec. 36(a)]. Some branch with ν vertices of degree 4 has a total of $3\nu + 1$ nodes [Sec. 36(a)]. Thus, we have for a branch originating in the center

$$3\nu + 1 < (3n + 2)/2, \quad \nu < \frac{n}{2}.$$

In other words, the number of vertices of degree 4 in a branch from the center is at most m [cf. (2.49)]. The center is of degree 4, hence: the number of vertices of degree 4 of all the branches originating in the center is equal to $n - 1$.

We examine first the topological congruence, the number ρ_n' of topologically different free centric C-H trees with n vertices of degree 4. According to the preceding remark, ρ_n' is the number of nonequivalent configurations, modulo $\mathcal{S}_4$, of four planted C-H trees which contain a total of $n - 1$ nodes and none of which contains more than m nodes. The generating function of these planted trees is

$$R_0 + R_1 x + R_2 x^2 + \cdots + R_m x^m = r(x)$$

(notation of Sec. 50). Substituting this series, according to the theorem of the first chapter (Sec. 16), in the cycle index of $\mathcal{S}_4$ (cf. Sec. 21) and determining the coefficient of x^{n-1}, we find

(2.50) $$\rho_n' = \text{Coeff}_n \left\{ x \frac{\overset{m}{r(x)^4} + 6\overset{m}{r(x)^2}\overset{m}{r(x^2)} + 3\overset{m}{r(x^2)^2} + 8\overset{m}{r(x)}\overset{m}{r(x^3)} + 6\overset{m}{r(x^4)}}{24} \right\}.$$

For the spatial congruence, that is, with group A_4 and its cycle index (cf. Sec. 21), we find

(2.51) $\sigma_n' = \text{Coeff}_n \left\{ x \; \dfrac{\overset{m}{s}(x)^4 + 3\overset{m}{s}(x^2)^2 + 8\overset{m}{s}(x)\overset{m}{s}(x^3)}{12} \right\}.$

Remembering the alternatives of a symmetric or an asymmetric center, we obtain similarly as in Sec. 42, based on the formulas $(2\ \mathit{S}_4 - \mathit{A}_4)$ and $(\mathit{A}_4 - \mathit{S}_4)$ of Sec. 23,

(2.52) $\rho_{n0}' + \rho_{n1}'y + \rho_{n2}'y^2 + \cdots =$

$= \text{Coeff}_n \left\{ x \; \dfrac{\overset{m}{\varphi_1^2}\,\overset{m}{\varphi_2} + \overset{m}{\varphi_4}}{2} + xy \; \dfrac{\overset{m}{\varphi_1^4} - 6\overset{m}{\varphi_1^2}\overset{m}{\varphi_2} + 3\overset{m}{\varphi_2^2} + 8\overset{m}{\varphi_1}\overset{m}{\varphi_3} - 6\overset{m}{\varphi_4}}{24} \right\}$

where

$$\sum_{v=0}^{m} \sum_{\alpha=0}^{\infty} R_{v\alpha} x^{vk} y^{\alpha k} = \overset{m}{\varphi_k}.$$

Setting $y = 0$ in (2.52) we get

(2.53) $\rho_{n0} = \text{Coeff}_n \left\{ x \; \dfrac{\overset{m}{r}(x)\,\overset{m}{r}(x^2) + \overset{m}{r}(x^4)}{2} \right\}.$

52. Now we get to the determination of the number τ_n' of topologically different free centric trees with n vertices. Select those among the τ_n' trees whose center carries s branches. The number of such trees is, according to the preceding discussion, equal to the number of nonequivalent configurations of s planted trees with respect to S_3. It is the coefficient of x^{n-1} in the expression we obtain by replacing $f(x,y,z)$ in (1.12) by

$$T_1 x + T_2 x^2 + \cdots + T_m x^m = \overset{m}{t}(x).$$

Denote the result of this calculation by

$$\text{Coeff}_n\{x\ F_s\}.$$

Putting successively $s = 0, 1, 2, \dots$, we obtain

(2.54) $\tau_n' = \text{Coeff}_n\{x(1 + F_1 + F_2 + F_3 + \cdots)\}.$

Specialize (1.11) in two ways: Firstly, let

$$f(x,y,z) = \overset{m}{t}(x),$$

and correspondingly

$$a_{k00} = T_k \quad \text{for} \ 1 \leqslant k \leqslant m,$$

$a_{k\lambda\mu} = 0$ if any of the three conditions

$k = 0, k > m, \quad \lambda + \mu > 0$

is satisfied. Secondly, let $u = 1$. Comparison of the first two lines
of (1.11) and (2.54) leads to

(2.55) $\tau_n' = \text{Coeff}_n \left\{ x(1 - x)^{-T_1}(1 - x^2)^{-T_2} \dots (1 - x^m)^{-T_m} \right\}.$

Remarks on Computations

53. Consider, as an example, the functional equation (4) which is
satisfied by the power series (3). Comparing the coefficients we
find the equations

$R_0 = 1,$

(2.56) $R_1 = (R_0^3 + 3R_0^2 + 2R_0)/6,$

$R_2 = (R_0^2 R_1 + R_0 R_1)/2,$

$\cdot \quad \cdot \quad \cdot \quad \cdot \quad \cdot \quad \cdot \quad \cdot \quad \cdot \quad \cdot \quad \cdot \quad \cdot,$

and, generally, R_n as a polynomial in $R_0, R_1, \dots, R_{n-1}$. Thus, we can
determine R_n recursively; Q_n, S_n, T_n can similarly be determined by
means of the functional equation. The computation of $R_{n\alpha}$ by means
of (2.22) is more cumbersome, involving a two-fold recursion.

The method to derive T_n was given by Cayley,[1] who established
the functional equation in the form of (1) for the function $t(x)$.
Cayley's computations of R_n are more laborious. Henze and Blair[2]
have derived the recursion formula (2.56) by direct combinatorial
considerations, without knowledge of the functional equation. Here,
(2.56) is a consequence of the functional equation (4).

The number ρ_n can be derived directly from (2.46) and (2.50) because of (2.41); however, the recursively established $R_0, R_1, \dots, R_m$,
and $R_{n/2}$ (in the case of n even), are involved in the computation.
The calculation of the quantities $\tau_n, \sigma_n, \rho_{n\infty}$ and κ_n is similarly tied
to the numbers $T_n, S_n, R_{n\infty}$ and Q_n, respectively.

The expressions (2.44) and (2.55), useful in computations, have
been derived by Cayley.[3] It is of interest to note that Cayley first

---

[1]Cayley, 1.

[2]Blair and Henze, 1.

[3]Cayley, 7.

computed (what we call) ρ_n and τ_n in a very tedious way by means of a different definition of center. He found the elegant formula (2.55) only later. He does not seem to have gone beyond the original, cumbersome method of calculating ρ_n; the much more useful approach which corresponds to formula (2.50) has been developed by Henze and Blair.[1]

The functional equations (1'), (4), (7), (8), (2.22), which have been established earlier[2] and proved in the present paper, not only summarize the recursion formulas for the numbers T_n, R_n, S_n, Q_n, $R_{n\infty}$ but allow also general inferences (e.g., Sec. 60), in particular on the asymptotic behavior (in Chapter 4).

Remarks on the Group of Automorphism of a Free Topological Tree

54. In this section a "tree" is understood to be a free tree with n vertices. Two trees are considered as distinct or nondistinct depending on whether they are topologically different or not. Thus, the number of different trees is τ_n,

$$\tau_n = \tau$$

for short.

The group of automorphisms of a tree comprises all one-to-one mappings of the tree onto itself which satisfy the conditions (I), (II), (III); that is, which map vertices onto vertices, edges onto edges while conserving the fundamental relationship. The group of automorphisms can be regarded as permutation group of the n points of the tree. Indeed, if each of the n points remain invariant under the automorphism, then the $(n-1)$ edges remain invariant. I insert two remarks on automorphisms which are loosely tied to various of the preceding discussions.[3]

(a) All $n!$ permutations of the symmetric group can be generated by $n-1$ suitable transpositions. A tree can be assigned to a collection of $n-1$ transpositions that generate $\mathcal{S}_n$. All collections which are conjugates of each other with respect to $\mathcal{S}_n$ belong to the same tree, and the normalizer of the collection is the group of automorphisms

---

[1] Blair and Henze, 2.

[2] Pólya 3, 4, 5. The last paper contains a direct proof for the equivalence of the recursion formulas based on combinatorial considerations and on the functional equation (4).

[3] König 1, p. 5, raises an interesting question about groups of graphs.

of the assigned tree. There are a total of n^{n-2} such collections.[1] Denoting by h_1, h_2, ..., h_T the orders of the groups of automorphisms which correspond to the τ different trees, we have

$$\frac{1}{h_1} + \frac{1}{h_2} + \cdots + \frac{1}{h_T} = \frac{n^{n-2}}{n!} \, .$$

This equation is stronger than the resulting inequality (2.40).

(b) Jordan[2] indicated a method to determine the order of the group of automorphisms of an arbitrary graph. For graphs of connectivity number 0, i.e., for trees, Jordan's procedure yields a more tangible result than for higher connectivity numbers, namely the following:

Each tree is associated with certain natural numbers m_1, m_2, ..., m_r

$$r \geqslant 1, \quad m_1 < m_2 < \cdots < m_r, \quad m_1 + m_2 + \cdots + m_r \leqslant n,$$

such that the group of automorphisms of the tree can be constructed from the symmetric groups $\mathfrak{S}_{m_1}$, $\mathfrak{S}_{m_2}$, ..., $\mathfrak{S}_{m_r}$ by repeated application of the two in Sec. 27 discussed operations: the direct product $\mathfrak{G} \times \mathfrak{H}$ and the corona construction of $\mathfrak{G}$ with respect to $\mathfrak{H}$, $\mathfrak{G}[\mathfrak{H}]$. In particular, the order of the group of automorphisms has to assume the form

$$m_1!^{\alpha_1} \, m_2!^{\alpha_2} \, ... \, m_r!^{\alpha_r} \, ,$$

where α_1, α_2, ..., α_r are some natural numbers. On the basis of obvious examples, we find further: an integer can represent the order of a group of automorphisms of a tree if and only if it is of the form

$$1^{d_1} \, 2^{d_2} \, ... \, m^{d_m} \, ,$$

where d_1, d_2, ..., d_m are natural numbers and $d_1 \geqslant d_2 \geqslant \cdots \geqslant d_m \geqslant 1$. That means, to each number

$$1, 2, 4, 6, 8, 12, 16, \, ...$$

there exists a tree the order of whose group of automorphisms is equal to that number, while no trees exist with a group of automorphisms of order

$$3, 5, 7, 11, 13, 14, 15, 17, 18, 19, 20, \, ... \, .$$

---

[1]Cf. footnote 2, p. 46.

[2]Jordan, 1.

On the other hand, every natural number is the order of the automorphism group of a graph with connectivity number 1.

Chapter 3
CHEMICAL COMPOUNDS

General Remarks

55. The elements of a graph have their interpretation in chemistry, the vertices are atoms, the edges are bonds, the graph turns into a chemical (structural) formula. Conditions I and II in Sec. 29 become meaningful in chemical terms. Every edge terminating in two end-points means that there are no free valences. The connectedness of a graph indicates that all atoms are tied together into a molecule. The number of edges ending in the same vertex corresponds to the valence of the atom: endpoints are atoms of valence one, vertices of degree k represent atoms of valence k.

In particular, a C-H graph represents a molecule of a chemical which is formed by atoms of valences 1 and 4. Unless the vertices are *a priori* required to be different, all the points of degree 4 present atoms of valence 4 and all the endpoints atoms of valence 1. By identifying the atom of valence 4 with C and the atom of valence 1 with H we turn the C-H graph into the formula of a carbohydrate. In particular, a free C-H tree with n vertices of degree 4, which must (Sec. 36) have $2n + 2$ endpoints, represents a paraffin C_nH_{2n+2}. A planted C-H tree with n edges of degree 4, a root, and $2n + 1$ endpoints is the structural formula of a mono-substituted paraffin, e.g., of $C_nH_{2n+1}Cl$. It is most natural to interpret a C-graph (cf. construction in Sec. 37) as carbon skeleton of a carbohydrate (for a mono-substituted paraffin).

Topological congruence is meaningful for arbitrary graphs which represent chemical formulas. Conditions (I), (II), (III) state explicitly what is automatically understood when one encounters a chemical formula: (I) indicates that the length or shape of the bonds is irrelevant, that only the presence or absence of a bond matters; (II) tells us that atoms of the same element cannot be distinguished and that atoms of different elements have to be distinguished; (I), (II), and (III) together point out that the relationships among the elements [i.e., the structure] are crucial. Since (IV) is meaningless for the topological congruence, among the elements, the conditions (I), (II), and (III) are sufficient in this case: the relationships, the structure, are all that matters. The topological congruence of graphs implies interpretation of a chemical formula as structure formula. I use the word structural isomers in this sense. For example, the number of different structural isomers C_nH_{2n+2} is equal to the number of topologically different free C-H trees with n vertices of degree 4; this number was called ρ_n in Sec. 36.

Spatial congruence of C-H graphs is applied essentially only in chemical formulas which represent a compound of carbon atoms and atoms of valence 1 (or radicals of valence 1). In this case condition (IV), besides (I), (II), (III), adds another restriction: not only the relationships are important but also the spatial arrangement of the bonds. The spatial interpretation of the congruence of C-H graphs coincides with the interpretation of the chemical formula as stereoformula. I use stereoisomers in this sense. For example, the number of different stereoisomers C_nH_{2n+2} is equal to the number of spatially different free C-H trees with n vertices of degree 4. In Sec. 36, this number was denoted by σ_n.

It is easy to establish the correspondence of the chemical interpretation of the numbers R_n, S_n, κ_n, and Q_n given in the Introduction, Sec. 4, and the definitions in graph theory discussed in the previous chapter (Sec. 36).

Two pictures of two spatial (three-dimensional) models can represent the same structural formula without representing the same stereoformula: they describe the same structural formula if they exhibit the same relationships (if they are topologically congruent, i.e., they satisfy conditions (I), (II), (III)). In order to describe the same stereoformula they must display the same relationships and the same spatial orientation [they satisfy (I), (II), (III), and in addition (IV) (with A_4), that is, be spatially congruent]. If two formulas viewed as stereoformulas are equal then they are certainly equal when they are treated as structural formulas. Consequently there are at least as many stereoisomers as there are structural isomers. This fact is reflected by (2.8). It is true particularly for paraffins and monosubstituted paraffins.

In this paper, asymmetric carbon atoms are considered only in paraffins and substituted paraffins, and the following definition will be retained (cf. Sec. 36(b)): A carbon atom is called asymmetric if the four bonded radicals are pairwise structurally different. (Thus, it is not sufficient to require that the four radicals are not stereoisomers in order to declare a carbon atom asymmetric. One could envisage other, possibly useful, definitions.)

Not all chemistry textbooks provide a clear description of what stereo- and structural formulas imply. Maybe the concepts of spatial and topological congruence of graphs could contribute to the clarification of the chemical terminology.

56.   Lunn and Senior[1] remarked that each basic chemical compound is associated with three permutation groups. The expositions of the first chapter of this paper gain considerably in importance in relation to the questions arising in chemistry.   Thus, the three groups are illustrated by cyclopropane, C_3H_6. This compound also provides   an   example   for   the   concepts   of   structural   and stereoformulas, of topological and spatial congruence.

The graph of cyclopropane (cf. Fig. 4) consists of three vertices of degree 4 which are joined by single bonds and of six endpoints which separate into three pairs.   The points of a pair are linked by single bonds to the same vertex of valence 4.   We are examining the question: In how many ways can the graph of cyclopropane be mapped congruently onto itself with respect to spatial and topological arrangements?

(a) *Spatial interpretation.* The triangle, whose vertices are the nodes of degree 4 allows 3! mappings on itself.   If a mapping of the triangle is selected, the mapping of the other elements of the graph is fixed.   Indeed, the mapping of the triangle determines the mapping of two edges of each of the coronas of the three vertices of degree 4.   This fixes, however, the mapping of the other two edges of the corona since the group A_4 is twice transitive.   (Intuitively: If you fix two vertices and the center of a rigid tetrahedron, the tetrahedron cannot be rotated, the other two vertices are held fast.)   The number of spatially congruent selfmaps is 6.

(b) *Topological interpretations.* As before, the triangle whose vertices are the points of degree 4 can be mapped onto itself in 6 ways. Once a mapping is chosen, the remaining elements of the graph can be permutated in $2^3 = 8$ ways.   Indeed, the two endpoints

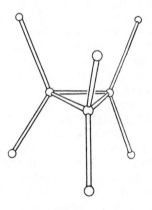

Figure 4

[1]Lunn and Senior.

which are directly linked to the same vertex of degree 4 can be interchanged. The number of topologically congruent selfmaps is, thus, $6 \times 8 = 48$.

The six selfmaps of (a) form a group. Elementary geometry provides a simple interpretation of the selfmaps as the group of rotations which leave a right prism with regular triangular base invariant. In fact, the 6 endpoints of the graph (see Fig. 4) can be matched with the six vertices of the prism. The six endpoints are subjected to a permutation group which we may call the group of the stereoformula. Its cycle index (Sec. 10) is

(3.1) $(f_1^6 + 3f_2^3 + 2f_3^2)/6.$

The 48 selfmaps of (b) form a group which is a permutation group of the six vertices. [The permutation group is the same as the permutation group of the six vertices of an octahedron subject to the 48 rotations and reflections which leave the octahedron invariant. This group is best described in the notation of Sec. 27 as $\mathcal{S}_3[\mathcal{S}_2]$: $\mathcal{S}_3$ interchanges the three C atoms (the three vertices of the triangle in Fig. 4) corresponding to the three diagonals of the octahedron, and $\mathcal{S}_2$ the two H-atoms linked to a C-atom corresponding to the two endpoints of an octahedral diagonal.] We may call this permutation group of order 48 the group of the structural formula. Its cycle index is best derived as an example of the formula (1.40), cf. also ($\mathcal{S}_2$) and ($\mathcal{S}_3$) in Sec. 21:

(3.2) $\left\{ \left[\dfrac{f_1^2 + f_2}{2} \right]^3 + 3\, \dfrac{(f_1^2 + f_2)\,(f_2^2 + f_4)}{22} + 2\, \dfrac{f_3^2 + f_6}{2} \right\} / 6.$

A third permutation group of the graph of cyclopropane obtains if the regular prism discussed as a model for (a) is subjected to rotations as well as to reflections which leave it invariant. The six vertices are thus subject to a permutation group of order 12. We call it the extended group of the stereoformula. Its cycle index is

(3.3) $(f_1^6 + f_2^3 + 3f_2^3 + 2f_3^2 + 3f_1^2 f_2^2 + 2f_6)/12.$

[One can establish that the 12 rotations which leave a regular hexagon invariant subject the six vertices to the 12 permutations of the extended group of the stereoformula of cyclopropane; the relationship between the Ladenburg prism formula and the usual hexagon formula of benzene is based on this. Formula (3.3) follows from ($\mathfrak{D}_s$) in Sec. 21 for $s = 6$.]

The three groups are related in the following way: The group of the stereoformula of order 6 is a subgroup of the extended group of order 12 which, in turn, is a subgroup of the group of the structural formula of order 48.

57. The three groups discussed in Sec. 56 are fundamental for the understanding of the isomers of the derivatives of cyclopropane, C_3H_6, where the six H atoms are replaced by radicals of valence 1.

Recalling the model of Sec. 56, imagine six radicals of valence 1 at the six endpoints of the graph of cyclopropane (Fig. 4); they form a configuration and each configuration provides the chemical formula of a derivative of cyclopropane. It is, however, possible that two different configurations, that is, different assignments of the radicals to the six endpoints, represent the same derivative. This is the case if and only if the two configurations can be transformed into each other by a permutation of the relevant group, that is, if they are equivalent with respect to the associated group. The relevant group is the group of the stereoformula in the case of stereoisomers, the associated group of the structural formula in the case of structure isomers. By extended group of the stereoformula (which we visualized as the group of 12 rotations and reflections of the right prism with a regular triangle as base) we understand the following: the permutations of the group transform the pairs of spatially different isomers, which are mirror images of each other (optical antipodes), into each other. Mirror image isomers are equivalent with respect to the extended group of the stereoformula, the two antipodes of a pair of mirror images are not distinct.[1]

Calculating the number of configurations which are nonequivalent with respect to the three groups we find:

for the group of the stereoformula the numbers of stereoisomers;

for the extended group of the stereoformula, the number of stereoisomers minus the number of pairs of mirror images;

for the group of the structural formula, the number of structure isomers.

As an example we calculate the number of different isomeric substitutes of cyclopropane of the form

$$C_3 \, X_k \, Y_\ell \, Z_m \, ,$$

where $k + \ell + m = 6$ and X, Y, Z are different independent radicals.[2]

---

[1]Stereo differences within the individual substitutes are not taken into account when we talk about antipodes.

[2]Independent means that $X_k \, Y_\ell \, Z_m$ and $X_{k'} \, Y_{\ell'} \, Z_{m'}$ have the same molecular structure only if $k = k'$, $\ell = \ell'$, $m = m'$. For example, the radicals -H, -CH_3, -C_2H_5 are not independent of each other because $C_3H_5 \, (C_2H_5)$ and $C_3H_4 \, (CH_3)_2$ have the same molecular structure. The discussion of the simultaneous substitution of different alkyls will be taken up in the next section.

According to the main theorem of Chapter 1, we introduce

$$f_1 = x + y + z, \quad f_2 = x^2 + y^2 + z^2, \quad f_3 = x^3 + y^3 + z^3$$

into the cycle index. Expansion of the series yields for the three groups, respectively,

$$x^6 + x^5 y + 4x^4 y^2 + 5x^4 yz + 4x^3 y^3 + 10x^3 y^2 z + 18x^2 y^2 z^2 + \cdots ,$$

$$x^6 + x^5 y + 3x^4 y^2 + 3x^4 yz + 3x^3 y^3 + 6x^3 y^2 z + 11x^2 y^2 z^2 + \cdots ,$$

$$x^6 + x^5 y + 2x^4 y^2 + 2x^4 yz + 2x^3 y^3 + 3x^3 y^2 z + 5x^2 y^2 z^2 + \cdots .$$

The coefficient of $x^4 y^2$ in the three expressions indicates that there exist four different derivatives of cyclopropane of the form $C_3 H_4 X_2$ (disubstituted cyclopropane with two identical substituents). Two of the four derivatives are mirror images of each other, that is, they form a pair of optical antipodes. If the spatial arrangement is disregarded, only two distinct cyclopropanes with formula $C_3 H_4 X_2$ are left, two structural isomers.

We note that chemical substitution of a radical into a basic compound corresponds (in the sense of the main theorem of Chapter 1) to the algebraic substitution of the generating function into the cycle index of the group of the basic compound.

58. The series $r(x)$, in which the coefficient of x is equal to R_n, the number of structurally isomeric alcohols $C_n H_{2n+1} OH$, is the generating function of these alcohols. However, it is also the generating function of the alkyl radicals $C_n H_{2n+1}$. Chemical substitution of the alkylradicals $C_n H_{2n+1}$ for H in the cyclopropane molecule $C_3 H_6$ leads to the homologues of cyclopropane. Algebraic substitution of the generating function $r(x)$ of the alkyl radicals in the cycle index (3.2) of the group of the structural formula of cyclopropane yields the generating function of structurally isomeric homologues of cyclopropane

$$(3.4) \qquad 1 + x + 3x^2 + 6x^3 + 15x^4 + 33x^5 + \cdots .$$

In detail, we find the series (3.4) according to the main theorem of Chapter 1 by setting

$$f_1 = r(x), \quad f_2 = r(x^2), \quad f_3 = r(x^3)$$

in the cycle index (3.2) and expanding in powers of x. The coefficient of x^n in the series (3.4) equals the number of structurally isomeric homologues with molecular formula $C_{3+n} H_{6+2n}$. We determine the number of stereoisomeric homologues of cyclopropane with the same molecular formula by introducing the generating function

$s(x)$ of the stereoisomeric alkyl radicals in the cycle index (3.1) of cyclopropane.

59. The procedure used in the preceding sections for cyclopropane serves equally well in the analytic determination of the numbers of structure and stereoisomeric compounds which obtain when essentially different radicals of valence one or alkyl radicals are substituted in the basic compound. We have to assume, however, that there is enough information on the basic compound to determine the three groups discussed in Sec. 56. This is certainly the case for the most important basic compounds, benzene and naphthalene. I omit the formulation of rules[1] which are obvious in the preceding example.

The example demonstrates that the concepts in chemistry rely heavily on notions from group theory, specifically the concept, introduced in Sec. 11, of the equivalence of configurations with respect to a permutation group. The cycle index and the main theorem of Sec. 16 play a role.

Referring again to the paper of Lunn and Senior, I conclude these general remarks and turn to the analytic determination of the number of isomers of certain special compounds.

Special Questions

60. *Structural isomers of* $C_nH_{2n+1}OH$ *with a given number of asymmetric C-atoms.* We return to the number $R_{n\alpha}$ which has been defined in Sec. 42 with respect to graphs.

$R_{n\alpha}$ denotes the number of distinct structural isomers of $C_nH_{2n+1}OH$ with exactly α asymmetric carbon atoms.[2]

The generating function $\phi(x,y)$ of $R_{n\alpha}$ and its functional equation (2.22) have been established in Sec. 42. Now we derive some properties of the numbers $R_{n\alpha}$ from the functional equation (2.22).

(a) *Determination of the lowest* $C_nH_{2n+1}OH$ *with a given number of asymmetric C-atoms.* The molecule $C_nH_{2n+1}OH$ contains n carbon atoms, thus there cannot be more than n asymmetric C-atoms,

$$R_{n\alpha} = 0 \quad \text{for} \quad \alpha > n.$$

This trivial remark prompts the question: For which combinations of α and n does $R_{n\alpha}$ vanish; for which is it positive?

Since $\phi(0,0) = 1$ and since the last term on the right-hand side of the functional equation (2.22) has non-negative coefficients (cf. Sec.

---

[1]Cf. Sec. 77 and Pólya 4.

[2]Cf. definitions in Sec. 36(b).

23, final statement) the left-hand side dominates the right-hand side,

$$\phi(x,y) \geqslant x\phi(x,y),$$

which implies

$$R_{n,\alpha} \geqslant R_{n-1,\alpha} \,.$$

Thus we can reformulate the question: Which is the first non-vanishing term in the monotone sequence $R_{0\alpha}$, $R_{1\alpha}$, $R_{2\alpha}$, ... ?
  Put

$$(3.5) \qquad R_{0\alpha} + R_{1\alpha}x + \cdots + R_{n\alpha}x^n + \cdots = q^{[\alpha]}(x).$$

We want to determine the first nonvanishing term of this power series.
  Put

$$q^{[0]}(x) = q(x), \quad q^{[1]}(x) = q^{\mathrm{I}}(x), \quad q^{[2]}(x) = q^{\mathrm{II}}(x), \,...$$

(the first equation coincides with (6) and (2.19)), and rewrite $\phi(x,y)$ in this notation:

$$(3.6) \qquad \phi(x,y) = q(x) + q^{\mathrm{I}}(x)y + q^{\mathrm{II}}(x)y^2 + q^{\mathrm{III}}(x)y^3 + \cdots .$$

Introduce the variable

$$(3.7) \qquad x^2 y = z$$

and define the function $\Psi(x,z)$ by

$$(3.8) \qquad \phi(x,y) = 1 + x + x^2\Psi(x,z).$$

Substituting (3.8) in (2.22) we find the functional equation for $\Psi(x,y)$:

$$\Psi(x,z) = 1 + z\Psi(x,z) + x\, P(x,\, z,\, \Psi(x,z),\, \Psi(x^2,z^2),\, \Psi(x^3,z^3)),$$

where P stands for a specific polynomial in five variables. The structure of the functional equation makes it clear that $\Psi(x,z)$ is a power series in x and z; that is, none of the exponents in the expansion of Ψ is negative. In particular, it satisfies the two relations

$$\Psi(0,z) = 1 + z\Psi(0,z),$$

$$\Psi(0,z) = 1 + z + z^2 + z^3 + \cdots .$$

Comparing the above to the expansion derived from (3.6), (3.7), and (3.8),

$$\Psi(x,z) = \frac{q(x)-1-x}{x^2} + \frac{q^{\mathrm{I}}(x)}{x^4}z + \frac{q^{\mathrm{II}}(x)}{x^6}z^2 + \cdots + \frac{q^{[\alpha]}(x)}{x^{2\alpha+2}}z^\alpha + \cdots ,$$

we find

(3.9)
$$R_{0\alpha} = R_{1\alpha} = \cdots = R_{2\alpha+1,\alpha} = 0, \quad R_{2\alpha+2,\alpha} = 1,$$

$$R_{n\alpha} \geq 1 \quad \text{for} \quad n \geq 2\alpha + 2.$$

The result contains the statement: *An alcohol* $C_nH_{2n+1}OH$ *containing* α *asymmetric C-atoms must have at least* $2\alpha + 2$ *carbon atoms. If the number of C-atoms is* $n = 2\alpha + 2$, *there exists exactly one structure* $C_nH_{2n+1}OH$ *with* α *asymmetric C-atoms.*

(b) *Determination of the lowest* $C_nH_{2n+1}OH$ *in which the asymmetries compensate each other.* We have mentioned in Sec. 42 that

$$\phi(x,0) = q(x), \quad \phi(x,1) = r(x).$$

What is the relation between $\phi(x,2)$ *and* $s(x)$? Equation (2.22) implies

$$\phi(x,2) = 1 + x(\phi(x,2)^3 + 2\phi(x^3,8))/3$$

or

(3.10)
$$\phi(x,2) = 1 + x(\phi(x,2)^3 + 2\phi(x^3,2))/3$$
$$+ 2x(\phi(x^3,8) - \phi(x^3,2))/3.$$

Compare the second equation to (7):

$$s(x) = 1 + x(s(x)^3 + 2s(x^2))/3.$$

According to (3.6) the power series

$$\phi(x,8) - \phi(x,2) = (8-2)q^{\mathrm{I}}(x) - (64-4)q^{\mathrm{II}}(x) + \cdots = 6x^4 + \cdots$$

has no negative coefficients; the first positive coefficient was derived from (3.9). Therefore, the power series

(3.11) $$\frac{2x}{3}(\phi(x^3,8) - \phi(x^3,2)) = 4x^{13} + \cdots$$

has only non-negative coefficients.

Comparison of (7), (3.10), and (3.11) shows that $s(x)$ is dominated by $\phi(x,2)$ (by arguments similar to those given below in Sec. 68); that is, taking the definition of these series into account (cf. (5), (2.20), respectively), we have the inequality

(3.12) $$S_n \leq R_{n0} + 2R_{n1} + 4R_{n2} + 8R_{n3} + \cdots .$$

Careful inspection of the result [recall the first term of (3.11)] reveals: The equal sign in (3.12) holds for $n \leqslant 12$, and strict inequality for $n \geqslant 13$.

Inequality (3.12) ensues from the well-known fact that a given structure which contains α asymmetric carbon atoms gives rise to 2^α, in general distinct, stereoisomers and in some exceptional cases to fewer than 2^α stereoisomers. Nevertheless, the purely analytical deduction of inequality (3.12) from (7) and (2.22) corroborates the observation. The exception, that is the case in which there are fewer than 2^α stereoisomers in the presence of α asymmetric carbon atoms, involves compensation of asymmetries. The corollary of (3.12) indicates that compensation of asymmetries in $C_n H_{2n+1} OH$ cannot occur for $n \leqslant 12$, that it does occur, however, for $n \geqslant 13$ for each n in at least one structure formula. [For $n = 13$, compensation occurs in exactly one structure formula, as can be seen by checking the numerical value of the first coefficient of (3.11) and by scrutinizing the chemical formula involved. The compound is described by $(C_4 H_9)_3 COH$, where $C_4 H_9$ is shorthand for

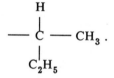

The number of asymmetric carbon atoms in this formula is $\alpha = 3$; the number of distinct stereoisomers is 4, not 8; the difference $8 - 4$ appears in (3.1) as the first coefficient.]

61. *Structurally isomeric disubstituted paraffins.* The number R_n of structurally isomeric alcohols $C_n H_{2n+1} OH$ is, apparently, the same as the number of any monodistributed paraffins $C_n H_{2n+1} X$, where X is a given radical of valence one (but not an alkyl); e.g., -Cl, -Br, -OH, etc.

Now we focus on the number of different structurally isomeric disubstituted paraffins of the form $C_n H_{2n} XY$, where X and Y are radicals of valence one which are distinct from each other and from alkyls (e.g., $X = $ -OH, $Y = $ -Cl). We aim directly at the generating function in which the coefficient of x^n is equal to the desired number.

The structure formula of $C_n H_{2n} XY$ is a tree with n vertices of degree 4 and $2n + 2$ endpoints. These are partitioned into three types, $2n$ endpoints are hydrogen atoms, one is an X and one a Y. Consider the path from X to Y in this tree. Let m be the number of C-atoms on this path.

If $m = 0$, the number of carbon atoms is $n = 0$, and we are dealing with the (formal) compound XY, which we include.

If $m = 1$ we are dealing with a derivative of $CXYH_2$ (disubstituted methane) which obtains when two H-atoms are replaced by alkyl radicals. The two H-atoms are exchangeable (structural, not stereo-isomers!), its group is the symmetric group $\mathcal{S}_2$ whose cycle index is given in Sec. 21. Replacing $r(x)$ in Sec. 21 ($\mathcal{S}_2$) and taking the C-atom in $CXYH_2$ into account we get the generating function of the special disubstituted paraffins which correspond to the condition $m = 1$,

(3.13) $x(r(x)^2 + r(x^2))/2 = xR(x)$.

(The abbreviation $R(x)$ will be used repeatedly.)

If $m > 1$, i.e., if X and Y are not bound to the same C, we are dealing with a derivative of a normal paraffin in which the H-atoms at the two ends are substituted, i.e. with

$$CH_2X\ (CH_2)_{m-2}\ CH_2Y.$$

The derivative obtains when the m pairs of H-atoms are replaced by alkyl radicals. The group of the structure formula (Sec. 56) is the direct product (Sec. 27) of m factors

$$\mathcal{S}_2 \times \mathcal{S}_2 \times \cdots \times \mathcal{S}_2.$$

Because X and Y are different, no pairs of -H are interchangeable. The cycle index is, according to Sec. 27,

$$[(f_1^2 + f_2)/2]^m.$$

Replacing the H-atoms by alkyl radicals, that is, replacing f by the generating series $r(x)$ of the alkyl radicals (as in Sec. 58) and representing the m C-atoms of the initial compound by the factor x^m, we get the generating function of the special disubstituted paraffins discussed here, namely

(3.14) $x^m((r(x)^2 + r(x^2))/2)^m = [xR(x)]^m$.

The sum of (3.14) over all m, $m = 0, 1, 2, ...$, equals

(3.15) $1/(1 - xR(x))$.

It is the generating function of all structurally isomeric disubstituted paraffins $C_nH_{2n}XY$.

62. *Trisubstituted paraffins.* A trisubstituted paraffin with three different substituents X, Y, Z (none of which can be an alkyl radical) has the molecular formula $C_nH_{2n-1}XYZ$. The structure formula is a tree. Examine the three linking paths XY, YZ, and ZX. It is easy to see that there exists one and only one point which is common to all three paths, a vertex of degree 4 (a C-atom); we call it "traffic center" of the tree, point V for short (V for "Verkehrszentrum").

The tree can be constructed in five steps and, correspondingly, the generating function of the structurally isomeric $C_nH_{2n-1}XYZ$-molecules can be built up from five factors (multiplication of the generating function in the case of independence; cf. Sec. 17).

We start with the point V, which corresponds to the factor x (a C-atom). Next, V is connected to X by a path which contains m carbon atoms which are capped by a pair of alkyl radicals, $m = 0, 1, 2, \dots$. This construction produces the factor (3.15), as we have seen in the previous section.

The construction is now repeated for V and Y, generating another factor (3.15). The connection of V with Z gives rise to a third factor (3.15). Finally, V is linked to an alkyl radical, which introduces the factor $r(x)$.

The generating function of the structural isomers $C_nH_{2n-1}XYZ$ is

$$(3.16) \qquad \frac{xr(x)}{(1 - xR(x))^3} .$$

63. *Multiply substituted paraffins.* If there are more than three substitutions in a paraffin the generating function of the structural isomers becomes more involved. I will describe the construction but will forego the details of the proof.

The structural form of an ℓ-times substituted paraffin, with ℓ distinct substituents, that is, of $C_nH_{2n+2-\ell}X'X'' \dots X^{(\ell)}$, is a tree ℓ of whose $2n + 2$ endpoints are marked. We subject the tree to the following two operations as often as possible.

(a) Omission of an unmarked endpoint including its bond.
(b) Omission of a point of degree 2 and fusion of the two adjacent bonds.

We are left with a reduced tree with exactly ℓ distinct endpoints labeled X', X'', ..., $X^{(\ell)}$ and which contains, besides the endpoints, only vertices of degrees 3 and 4.

As in Sec. 30, p_1, p_3, p_4 denote the number of vertices of degree 1, 3, and 4, respectively, and s the number of its edges.

We have

$$p_1 = \ell$$

and the relations [cf. (2.1), (2.2), (2.3); note that $\mu = 0$]

$$\ell + p_3 + p_4 = s + 1, \qquad \ell + 3p_3 + 4p_4 = 2s$$

which combine to

$$(3.17) \qquad \begin{aligned} s &= 2\ell - 3 - p_4 \leqslant 2\ell - 3 \\ p_3 + p_4 &= s + 1 - \ell \leqslant \ell - 2. \end{aligned}$$

Hence, s, p_3, p_4 can assume only finitely many values: *For given ℓ there exist only finitely many topologically different reduced trees.* The following rule holds: *The generating function of the structurally isomeric $C_n H_{2n+2-\ell} X' X'' \dots X^{(\ell)}$ molecules is equal to*

$$(3.18) \qquad \Sigma x^{p_2+p_4} r(x)^{p_3} / [1 - xR(x)]^s;$$

the sum extends over all topologically different reduced trees which belong to the given value ℓ.

The simplicity of the formulas derived for $\ell = 2$ and $\ell = 3$ in the preceding two sections is due to the fact that in those two cases there exists only one reduced tree. The sum (3.18) consists, thus, in those two cases in a single term; we get

$$(3.15) \qquad \text{for } \ell = 2, \quad p_3 = 0, \quad p_4 = 0, \quad s = 1,$$

$$(3.16) \qquad \text{for } \ell = 3, \quad p_3 = 1, \quad p_4 = 0, \quad s = 3.$$

For the sake of completeness, we note: *The number of those reduced trees for which relation (3.17) is an equality, that is, the number of those topologically different trees with ℓ distinct endpoints and which have, besides the endpoints, only vertices of degrees 3 and 4, is equal to*

$$(3.19) \qquad 1 \cdot 3 \cdot 5 \dots (2\ell - 5) = (2\ell - 4)! / [2^{\ell-2} \cdot (\ell - 2)!].$$

For $\ell = 2$ and $\ell = 3$, the value of (3.19) reduces to 1. Relation (3.19) can easily be proved by induction.[1]

64. *Structurally isomeric cycloalkanes $C_n H_{2n}$.* The structure formula of these compounds is a graph of connectivity number 1 which is subject to two conditions:

(1) All vertices which are not endpoints have degree 4 (C-H graph in the sense of Sec. 35).
(2) Two vertices are linked by at most one path (the double bonds

    | |
    C=C and thus the homologues of paraffins are excluded).
    | |

If we remove from such graphs an endpoint including the edge belonging to the endpoint and if we keep repeating this process as often as possible, we end up with a ring; that is, a connected graph of m

---

[1] In the preceding section we looked only at those multiply substituted paraffins in which all the substituents are distinct. The case where two or more substituents are equal can be treated too; however, the description and justification of the formulas become so awkward that I refer the reader to the generating function established elsewhere for two and three substituents. Cf. Pólya 4, p. 440.

points and m edges, each of which is bounded by two different vertices while each vertex is endpoint of two different edges, $m = 3$, 4, 5, ... [$m \neq 2$ because of condition II]. Such a ring is the carbon skeleton (C-graph in the sense of Secs. 35 and 37) of a purely ring-shaped cycloparaffin. By substituting an alkyl radical into a purely ring-shaped cycloparaffin we obtain a non-purely ring-shaped cycloparaffin; that is, these are the homologues of the purely ring-shaped cycloparaffins.

It is easily seen that for given m there exists only one ring-shaped C-graph with m vertices. The construction described in Sec. 37 defines a topologically unique C-H graph: For a given m there exists with respect to structure exactly one purely ring-shaped cycloparaffin. It is, in the simplest case $m = 3$, the above (Secs. 56 - 58) discussed cyclopropane.

Generalizing properly what we have discussed for the case $m = 3$, we can establish the group of the structure formula of the only purely ring-shaped $C_m H_{2m}$: it has degree $2m$ and order $2m \, 2^m$ and the group is $D_m[S_2]$ in the notation of Secs. 21 and 27. Therefore, the cycle index of the group of the structure formula of the purely ring-shaped $C_m H_{2m}$ [cf. Sec. 21, (D_s), (S_2); further (1.39), (1.40)] is

$$\frac{1}{2m} \sum_{k|m} \varphi(k) \left[\frac{f_k^2 + f_{2k}}{2} \right]^{m/k}$$

(3.20)

$$+ \begin{cases} \dfrac{1}{2} \dfrac{f_1^2 + f_2}{2} \left[\dfrac{f_2^2 + f_4}{2} \right]^{\mu-1} & \text{for } m = 2\mu - 1, \\[2em] \dfrac{1}{4} \left[\left(\dfrac{f_1^2 + f_2}{2} \right)^2 + \dfrac{f_2^2 + f_4}{2} \right] \cdot \left[\dfrac{f_1^2 + f_4}{2} \right]^{\mu-1} & \\ & \text{for } m = 2\mu \end{cases}$$

In order to set up the generating function of the homologues of the purely ring-shaped $C_m H_{2m}$ we substitute the generating function $r(x)$ of the structurally isomeric alkyl radicals for f according to Sec. 58, corresponding to the chemical substitution of the alkyl radical for -H. The m carbon atoms in $C_m H_{2m}$ contribute the factor x^m. Thus (3.20) becomes, with the abbreviation of (3.13),

(3.21) $\dfrac{1}{2m} \sum_{k|m} \varphi(k)[x^k R(x^k)]^{m/k} + \begin{cases} \dfrac{1}{2} x R(x)[x^2 R(x^2)]^{\mu-1} \\[1.5em] \dfrac{1}{4}\{[x R(x)]^2 + x^2 R(x^2)\} \cdot [x^2 R(x^2)]^{\mu-1}. \end{cases}$

Summing (3.21) over $m = 3, 4, 5, ...$, we obtain the generating function of the structural isomeric cycloparaffins which we will denote by P. Summation over $m = 1, 2, 3, ...$ is easier. Summation and recombination of the terms provide the following sequence of results:

$$xR(x) + \frac{x^2}{2} (R(x)^2 + R(x^2)) + P = \frac{1}{2} \sum_{m=1}^{\infty} \frac{1}{m} \sum_{k|m} \varphi(k)[x^k R(x^k)]^{m/k}$$

$$+ \left\{ \frac{x}{2} R(x) + \frac{x^2}{4}[R(x)^2 + R(x^2)] \right\} \sum_{\mu=1}^{\infty} [x^2 R(x^2)]^{\mu-1},$$

(3.22) $$P + \left\{ xR(x) + \frac{x^2}{2}(R(x^2) + R(x)^2) \right\} \left[1 - \frac{1}{2} \frac{1}{1 - x^2 R(x^2)} \right]$$

$$= \frac{1}{2} \sum_{m=1}^{\infty} \sum_{k\ell=m} \frac{\varphi(k)}{k\ell}\{x^k R(x^k)\}^\ell = \frac{1}{2} \sum_{1}^{\infty} \frac{\varphi(k)}{k} \sum_{\ell=1}^{\infty} \frac{\{x^k R(x^k)\}^\ell}{\ell}$$

$$= \frac{1}{2} \sum_{k=1}^{\infty} \frac{\varphi(k)}{k} \log \frac{1}{1 - x^k R(x^k)} .$$

This yields the power series

$$P = x^3 + 2x^4 + 5x^5 + 12x^6 + 29x^7 + 73x^8 + \cdots ,$$

in which the coefficient of x^n is the number of structurally isomeric alkenes $C_n H_{2n}$ (without double bonds).

65. *Hydrocarbons* $C_n H_{2n+2-2\mu}$. The structure formula of such a compound is a C-H graph of connectivity number μ [cf. Sec. 36(a)]. It is possible to write the generating function of the structurally isomeric $C_n H_{2n+2-2\mu}$ as has been done for $\mu = 1$. From an analytic point of view the function is simpler for $\mu \geqslant 2$ than for $\mu = 1$: It is a rational function of finitely many of the quantities x, $r(x)$, $r(x^2)$, $r(x^3)$, ... The formulas, though, are so unwieldy that I restrict myself to a few hints for the case $\mu = 2$.

Subject a graph of some $C_n H_{2n-2}$ to the two procedures described in Sec. 63(a) and (b) as often as possible. Since there are, in this case, no distinct endpoints the graph is eventually stripped of all its endpoints and we are left with one of the three in Fig. 5 represented forms. To fix ideas, consider, for example, the one in which each arc has two distinct endpoints. (This is, incidentally, the only one among the three forms which can be called a graph according to the definition of Sec. 29; don't forget requirement I!)

Partitioning the three arcs by k, ℓ, m points, respectively, that is, into $k + 1$, $\ell + 1$, $m + 1$ segments, we obtain the graph of a carbon skeleton (C-graph) of a certain basic compound $C_{k+\ell+m+2} H_{2k+2\ell+2m+2}$. One can determine the group of its structure formula; it has degree $2k + 2\ell + 2m + 2$ and order 12, 4, or 2, depending on $k = \ell = m$, $k = \ell \neq m$, or all

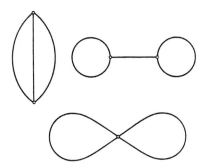

Figure 5

three numbers k, ℓ, m being distinct. One can establish the cycle index of this group and, by substituting $r(x)$ in the index, derive the generating function of the homologues, that is, of those hydrocarbons C_nH_{2n-2} which result when alkyl radicals are substituted in the basic compound. Summation over the triple k, ℓ, m leads to geometric series; I restrict my attention to the principal term,

(3.23) $$\frac{1}{12} \frac{x^2 r(x)^2}{(1 - xR(x))^3},$$

and forego discussion of the other terms. The expression (3.23) consists of two factors. The first, constant, factor is the reciprocal of the highest order the group of one of the basic compounds considered can achieve. The second, x-dependent factor is of the type of the summands in (3.18) with

$$p_3 = 2, \quad p_4 = 0, \quad s = 3,$$

and corresponds to independent assignments to the two points and the three arcs: alkyl radicals are placed at the two vertices; m C-atoms, each bonded to two alkyls, are placed on the arc. Summation extends over $m = 0, 1, 2, \dots$; cf. the derivation of (3.15) and (3.16).

The other two forms of Fig. 5 are tied to similar generating functions whose principal terms are

$$\frac{1}{8} \frac{x^2 r(x)^2}{(1 - xR(x))^3}, \quad \frac{1}{8} \frac{x}{(1 - xR(x))^2},$$

respectively. Both expressions correspond to the same group of order 8 and to the parameters

$$p_3 = 2, \quad p_4 = 0, \quad s = 3,$$

and

$$p_3 = 0, \quad p_4 = 1, \quad s = 2,$$

respectively. The notation "principal term" will be explained in Sec. 79.

Chapter 4
ASYMPTOTIC EVALUATION OF THE NUMBER OF COMBINATIONS

Function Theoretic Properties

66. In the preceding section, we have established the importance of the power series $q(x)$, $r(x)$, $s(x)$, $t(x)$ in combinatorics. Here we examine their analytical properties: radius of convergence, singularities on the circle of convergence, analytic continuation. We derive these characteristics from the functional equations whose solutions these series present. I start with a summary of the equations and some notations.

$$(4.1) \qquad f = 1 + xf = \sum_{0}^{\infty} x^n,$$

$$(4.2) \qquad q = f = 1 + xff_2,$$

$$(4.3) \qquad r = f = 1 + xff_2 + x(f^3 - 3ff_2 + 2f_3)/6,$$

$$(4.4) \qquad s = f = 1 + xff_2 + 2x(f^3 - 3ff_2 + 2f_3)/6,$$

$$(4.5) \qquad f = 1 + xf^3 = 1 + \sum_{1}^{\infty} \binom{3n}{n-1} \frac{x^n}{n},$$

$$(4.6) \qquad r - 1 = f = x\left[1 + f + \frac{f^2 + f_2}{2} + \frac{f^3 + 3ff_2 + 2f_3}{6}\right],$$

$$(4.7) \qquad t = f = x\left[1 + f + \frac{f^2 + f_2}{2} + \frac{f^3 + 3ff_2 + 2f_s}{6} + \cdots\right],$$

$$(4.8) \qquad f = xe^f = \sum_{n=1}^{\infty} \frac{n^{n-1}}{n!} x^n,$$

$$(4.9) \qquad t = f = xe^{f/1 + f_2/2 + f_3/3 + \cdots},$$

$$(4.10) \qquad f = xe^{f/1 + f^2/2 + f^3/3 + \cdots} = \frac{x}{1 - f} = \sum_{1}^{\infty} \binom{2n - 2}{n - 1} \frac{x^n}{n}.$$

The function which satisfies the equation is denoted by f, and $f_k = f(x^k)$, cf. (1.18), for short. The functional equation (4) of $r(x)$ appears twice, in different forms, as (4.3) and (4.6); similarly t is given by two different equations, (4.7) and (4.9); cf. (1'') and (1'). Equation (4.1) determines the geometric series, equations (4.5) and (4.10) determine algebraic functions, equation (4.8) the inverse of an elementary entire, transcendental function. The Maclaurin expansion of these four functions is easily derived, e.g., with the help of Lagrange's theorem.[1] For the combinatorial meaning of the series (4.5), (4.8), (4.10) see, for example, the discussions accompanying equations (2.15), (2.37), (2.27); with respect to (4.6), compare Sec. 44.

67. *Each of the functional equations* (4.1) - (4.10) *determines a unique power series. The constant term of the first five solutions is equal to* 1, *of the second five it is zero. The other coefficients are positive; moreover, they are integers except in the case of equation* (4.8). *The sequence of coefficients does not decrease.*

These statements are a consequence of the recursion relations obtained by identifying the coefficients of the power series expansion on the right- and left-hand side of the equation. For example, in (4.6), the coefficient of x^n is R_n ($n \geqslant 1$) on the left-hand side, and on the right-hand side a polynomial in $R_1, R_2, ..., R_{n-1}$ [cf. (2.56)], which implies the uniqueness. The coefficients of the polynomial mentioned are non-negative; the term R_{n-1} occurs, coming from xf, thus $R_n \geqslant R_{n-1}$; and so on. The statements that the coefficients are non-negative increasing integers are immediate consequences of the combinatorial considerations. Some assertions which concern the elementary functions follow from the explicitly given form of the coefficients.

The power series in x which is determined uniquely by the equation will be called the series associated with the equation.

68. *The ten equations* (4.1) - (4.10) *form three groups; the first consists of the first five, the second of the next two, and the third of the last three. Within a group they are ordered such that the succeeding power series is a majorant of the preceding power series.*

The seven inequalities of (10) and (12) in the Introduction are an expression of this statement. Meanwhile, the inequalities have been established with the help of combinatorial considerations [cf. in

---

[1]Bürmann-Lagrange theorem. Pólya-Szegö, *Problems and Theorems in Analysis*, Vol. I, (1972) pp. 145-146.

particular (2.7), (2.8), (2.16), (2.17); (2.9), (2.38), (2.39); (2.28), (2.29)].
The proof below of the inequality $R_n \leqslant S_n$ illustrates the derivation
by means of the recursion relationship.

We use the method of induction to prove the inequality. We have
$R_0 = S_0 = 1$ and assume that the inequality holds for all the terms,
up to $n - 1$ $(n \geqslant 1)$;

(4.11) $R_0 \leqslant S_0, \quad R_1 \leqslant S_1, \ ..., \ R_{n-1} \leqslant S_{n-1}.$

Let

$$f(x) = U_0 + U_1 x + U_2 x^2 + \cdots$$

denote a power series with indefinite coefficients $U_0, \ U_1, \ U_2, \ ...$.
Expand the two expressions

$$x(f(x)^3 + 3f(x) \ f(x^2) + 2f(x^3))/6,$$

$$x(f(x)^3 - 3f(x) \ f(x^2) + 2f(x^3))/6$$

in powers of x and denote the coefficient of x^n by $F(U_1, \ U_2, \ ..., \ U_{n-1})$,
$G(U_1, \ U_2, \ ..., \ U_{n-1})$, respectively. In this notation equations (4.3) and
(4.4) become

(4.12) $R_n = F(R_1, \ R_2, \ ..., \ R_{n-1}),$

(4.13) $S_n = F(S_1, \ S_2, \ ..., \ S_{n-1}) + G(S_1, \ S_2, \ ..., \ S_{n-1}).$

Both F and G are polynomials in the unknowns $U_0, \ U_1, \ ..., \ U_{n-1}.$
Obviously F has non-negative coefficients, hence, according to the
induction assumption (4.11) we have

(4.14) $F(R_1, \ R_2, \ ..., \ R_{n-1}) \leqslant F(S_1, \ ..., \ S_{n-1}).$

Some of the coefficients of G could be negative but G assumes non-
negative values for non-negative integers $U_0, \ U_1, \ ..., \ U_{n-1}$ (cf. final
observation of Sec. 23), and $S_0, \ S_1, \ ..., \ S_{n-1}$ are positive integers (cf.
Sec. 67). Therefore, the inequality

(4.15) $G(S_1, \ S_2, \ ..., \ S_{n-1}) \geqslant 0$

holds. Relations (4.12) - (4.15) imply

$$R_n \leqslant S_n.$$

69. The power series associated with the functional equations
(4.1) - (4.10) have all nonzero radii of convergence. For equations

$$\text{(4.1),} \quad \text{(4.5),} \quad \text{(4.8),} \quad \text{(4.10)}$$

the power series are explicitly given. The proposition is easily
verified. Inspection of the ratios of successive coefficients leads to
the respective radii of convergence

$$1, \quad \frac{4}{27}, \quad \frac{1}{e}, \quad \frac{1}{4}.$$

In the Introduction, the radii of convergence of the four power
series

$$q(x), \quad r(x), \quad s(x), \quad t(x)$$

have been called

$$\kappa, \quad \rho, \quad \sigma, \quad \tau.$$

In this case the statement follows from the inequalities (10) and (12),
which have been established in the preceding Sec. 68. To be exact,
we are not obtaining the strict inequalities (13) and (14) but the
weaker inequalities of equal or less, $\leqslant$ instead of $<$.

In all cases, (4.1) - (4.10), the power series associated with the
equation defines a function element with center at the origin. In
the sequel, we call it the function element associated with the equation. The function elements associated with the equations (4.1), (4.3),
(4.4), (4.7) are the power series $q(x)$, $r(x)$, $s(x)$, $t(x)$.

70. Examine first the analytic continuation of the function element given by $q(x)$.

The first inequality of (10) indicates that the radius κ of $q(x)$ is
at most equal to 1. It cannot be equal to 1: If $\kappa = 1$, then $q(x)$ and
consequently $q(x^2)$ converge in the unit circle and the power series

$$xq(x^2) = x + x^3 + \cdots$$

with all non-negative coefficients increases from 1 to above 2 as x
increases from $x = 0$ to $x = 1$ along the real axis. Thus, the function $xq(x^2)$ assumes the value 1 at an intermediate point ζ. Then the
function

$$(4.16) \qquad q(x) = \frac{1}{1 - xq(x^2)}$$

[relation (4.2)] has a pole at ζ, $0 < \zeta < 1$, contradicting the assumption that the radius of convergence is 1. Consequently, the radius of
convergence must be less than 1.

The power series of $q(x)$ converges in the circle $|x| < \kappa$, the power
series $q(x^2)$ converges in the circle $|x^2| < \kappa$, that is $|x| < \sqrt{\kappa}$; since $\kappa <$
1, we have $\kappa < \sqrt{\kappa}$, which means the region of convergence of $q(x^2)$
contains the region of convergence of $q(x)$. The singularities of $q(x)$
on the circle of convergence must be due to the zeros of the
denominator of (4.16), that is, they satisfy the equation

(4.17) $xq(x^2) = 1.$

The left-hand side is a power series with the coefficients of the even powers equal to zero and positive for the odd powers. The maximum absolute value of such a series on a circle centered at zero is assumed at exactly two points; on the positive and the negative real axis, and the sign of the function value is the sign of x. Therefore, equation (4.17) has a root on the positive real axis, specifically, a root closer to the origin than all the other roots. At this root the derivative of the left-hand side of (4.17) does not vanish, again as a consequence of the signs of the coefficients in the series for $xq(x^2)$, hence the root is simple. According to (4.16), this root determines the circle of convergence of $q(x)$.

To sum up: *The circle of convergence of $q(x)$ contains exactly one singular point, the point $x = \kappa$; $q(x)$ has a pole of first order at $x = \kappa$.*

71. The same argument implies that the analytic continuation of $q(x)$ in the unit circle is meromorphic. The argument is based on the functional equation (4.16) which is equivalent with the continued fraction (8'). The continuation of $q(x)$ is derived from the continuous fraction by a minor variation of the reasoning used above. We note the following shortcut.

Let

(4.18) $q(x) = \dfrac{\psi(x^2)}{\psi(x)}$.

Equation (4.2) turns into a functional equation for ψ,

(4.19) $\psi(x) = \psi(x^2) - x\psi(x^4),$

which is linear. The MacLaurin expansion of the solution,

$$\psi(x) = a_0 + a_1 x + a_2 x^2 + \cdots,$$

leads to the recursive relations

(4.20) $a_{2m} = a_m, \; a_{4m+1} = -a_m, \; a_{4m+3} = 0, \quad m = 0, 1, 2, ...,$

which uniquely determine the sequence of coefficients, if a_0 is given. Assume

(4.21) $a_0 = 1,$

then

$$\Psi(x) = 1 - x - x^2 - x^4 + x^5 - x^8 + x^9 + \cdots .$$

The coefficients can assume only the values 0, 1, −1, thus the series converges in the unit circle.

Relation (4.18) defines $q(x)$ in terms of $\psi(x)$. This function $q(x)$ is regular at $x = 0$ because of (4.21), and satisfies equation (4.2) because of (4.19). Since, according to Sec. 67, the solution of (4.2) in terms of a power series is unique, the function defined by (4.18) has to be identical to the former. The function $q(x)$ as represented by (4.18) is the quotient of two power series which converge in the unit disk, which implies that $q(x)$ is meromorphic in the unit disk.

We point out that numerator and denominator in the definition (4.18) of $q(x)$ have no zeros in common in the unit disk. Namely, it is impossible that there is a point x_0 such that

$$0 < |x_0| < 1, \quad \psi(x_0) = 0, \quad \psi(x_0^2) = 0.$$

If not, relation (4.19) would imply

$$\psi(x_0^4) = 0.$$

Repeating the argument we would find

$$\psi(x_0^8) = \psi(x_0^{16}) = \psi(x_0^{32}) = \cdots = 0,$$

that is, a sequence of zeros with accumulation point 0. This contradicts (4.21), thus $\psi(x)$ and $\psi(x^2)$ cannot have common zeros in the unit disk.

One can show with the help of the continued fraction (8') and arguments which are used in the theory of sequences of Sturm's fractions, that $q(x)$ has infinitely many poles between $x = \kappa$ and $x = 1$. The function $q(x)$ has the following properties:

firstly: $q(x) = Q_0 + Q_1 x + \cdots$ has integer coefficients;

secondly: $q(x)$ is meromorphic within the unit circle;

thirdly: $q(x)$ is not a rational function (e.g., because of the infinitely many poles; also by inspection of the degrees in (4.2)).

According to a general theorem,[1] the three properties combined imply that the unit circle is a singular curve for $q(x)$. This fact can be established without involving the general theorem, by making better use of the continued fraction (8'). A proof is outlined below.

72. This section deals with the analytic continuation of the function element $r(x)$.

The power series $r(x)$, $r(x^2)$, $r(x^3)$ converge within circles with center 0 and radii ρ, $\rho^{1/2}$, $\rho^{1/3}$, respectively. We note that by Secs. 68 and 70, $\rho \leqslant \kappa < 1$, hence

---

[1] F. Carlson, *Math. Zeitschrift*, 9(1921), pp. 1-13. G. Pólya, *Proc. London Math. Soc.* (2)21, (1921), pp. 22-38.

$$\rho < \rho^{1/2} < \rho^{1/3}.$$

Let $r(x) = y$ and consider the functional equation (4.3) in the form

(4.22) $xy^3 - 3[2 - xr(x^2)]y + 2[3 + xr(x^3)] = 0.$

The function $y = r(x)$ satisfies the equation (4.22) of degree 3 whose coefficients are regular on the disk $|x| < \rho^{1/2}$. Under analytic continuation of the function element $y = r(x)$ on this disk, singularities can arise in only two ways:

either the coefficient of the highest power in equation (4.22) vanishes; this occurs for $x = 0$ only;

or equation (4.22) has a multiple root y; in this case the partial derivative on the left hand side with respect to y vanishes, that is

(4.23) $xy^2 = 2 - xr(x^2).$

Elimination of y from (4.22) by means of (4.23) leads to the relation

(4.24) $[2 - xr(x^2)]^3 - x[3 + xr(x^3)]^2 = 0.$

Since the left-hand side of (4.24) is regular in the disk $|x| < \rho^{1/2}$, its zeros cannot have a point of accumulation, and in the neighborhood of a zero the branch of the function which becomes singular remains bounded and thus continuous.

We are interested in the singularities of the power series on its circle of convergence $|x| = \rho$, which is inside the disk $|x| < \rho^{1/2}$. The point $x = 0$ is not on the circle, thus the second of the above alternatives must hold: The singularities on the circle of convergence satisfy the equations (4.23) and (4.24), and the function element remains continuous. In particular, the power series $r(x)$ remains finite when x approaches ρ on the real axis. Since the coefficients of the power series are all non-negative it converges also for $x = \rho$ and consequently converges absolutely on the entire circle of convergence. Therefore, for the singular point x on the circle of convergence, equation (4.23) assumes the form

(4.25) $x(r(x)^2 + r(x^2))/2 = 1,$

where $r(x)$, $r(x^2)$ are power series.

Since the coefficients of the power series $r(x)$ are all positive, the point $x = \rho$ must be singular; thus, equation (4.25) holds for $x = \rho$. The coefficients of the power series on the left-hand side of (4.25) are all positive, except for the constant, which is equal to 0. On the circle of convergence $|x| = \rho$, the absolute value of the series assumes, therefore, its maximum at $x = \rho$. We conclude that $x = \rho$ is the only solution of (4.25) on the circle of convergence: *the point*

$x = \rho$ is the only singular point of $r(x)$ on the circle of convergence.

It is easy to see that the function element $r(x)$ can be expanded in powers of $\sqrt{x-\rho}$ and that the first terms of the expansion are

(4.26) $r(x) = a - b\sqrt{1 - x/\rho} + \cdots$,

where a and b are positive numbers:

(4.27) $a = r(\rho)$

(4.28) $b = \{2(r(\rho) - 1 + r(\rho)r'(\rho^2)\rho^3 + r'(\rho^3)\rho^4)/\rho r(\rho)\}^{1/2}.$

Examining the behavior of equation (4.3) beyond the circle of convergence $|x| = \rho^{1/2}$ we find that in any region of the open unit disk the continuation of $r(x)$ is algebroid, that is, it has only finitely many branches and finitely many algebraic branch points. The number of branches might increase to infinity if the region increases to the open disk.

Later we will need the following corollary: *On the closed disk* $|x| \leqslant \rho$, $x = \rho$ *is the unique solution of equation* (4.25).

73. The singularities of the functions $s(x)$ and $t(x)$ can be examined by the same methods. The equations corresponding to (4.16) and (4.22) are, respectively,

$$xy^3 - 3y + 2xs(x^3) + 3 = 0,$$

$$-y + xe^y \exp\left[\frac{t(x^2)}{2} + \frac{t(x^3)}{3} + \cdots\right] = 0.$$

The singularities on the circle of convergence are determined by the equations

(4.29) $xs(x)^2 = 1,$

(4.30) $t(x) = 1,$

respectively. They are analogous to the equations (4.17) and (4.25). Even on the circle of convergence, one can replace $s(x)$ and $t(x)$ [(4.29) and (4.30)] by their power series. Since the coefficients are all positive there is only one solution, namely the positive point on the circle of convergence: *the only singularities of the power series* $s(x)$, $t(x)$ *on their circles of convergence are* $x = \sigma$ *and* $x = \tau$, *respectively.* The function elements can be expanded near σ and τ in powers of $\sqrt{x-\sigma}$ and $\sqrt{x-\tau}$. Specifically we have

$$a' - b'\sqrt{1 - x/\sigma} + \cdots, \qquad a'' - b''\sqrt{1 - x/\tau} + \cdots,$$

where a', b', a'', b'', are positive.

74. We return to the series $r(x)$. Section 72 indicates that the radius of convergence ρ of $r(x)$ can be extracted from (4.24) or from (4.25). We use both possibilities.

(a) Equation (4.25) is, as discussed in Sec. 72, satisfied by $x = \rho$ and, because all the coefficients on the left-hand side are positive (except the constant term being 0), no value x with $0 < x < \rho$ satisfies (4.25). Thus, we may state the criterion: *a positive number x is smaller than ρ if and only if $r(x)$ converges and the left-hand side of (4.25) is smaller than 1.*

Examine the value $x = \sigma$; as implied in Sec. 73, the sequence $s(x)$ converges for $x = \sigma$ and it satisfies equation (4.29). Therefore the following equation holds:

(4.31) $\qquad \sigma s(\sigma)^2 = 1;$

we note that $r(\sigma)$ converges because of (10) and that

(4.32) $\qquad r(\sigma) \leqslant s(\sigma).$

Because the coefficients are positive and $r(0) = 1$, we have

(4.33) $\qquad r(\sigma)^2 > r(\sigma) > r(\sigma^2).$

The relations (4.31), (4.32), (4.33) imply

$$1 = \sigma s(\sigma)^2 \geqslant \sigma r(\sigma)^2 > \sigma(r(\sigma)^2 + r(\sigma^2))/2.$$

According to the criterion of the preceding paragraph we conclude $\sigma < \rho$.

All the inequalities of (13) and (14) in the Introduction can be derived by the same method.

(b) Equation (4.24) gives rise to the following criterion. *If x is a positive value for which the series $r(x^2)$ converges, then x is smaller or equal or larger than ρ, depending on whether*

$$\frac{x(3 + xr(x^3))^2}{(2 - xr(x^2))^3} < 1 \quad \text{or} = 1 \quad \text{or} > 1.$$

This criterion allows the computation of the radius of convergence ρ by means of convergent series. The numerical evaluation requires an estimate of the remainder terms of the series $r(x)$. Relations (10) and (12) provide the inequalities

(4.34) $\qquad R_n \leqslant T_n \leqslant \dfrac{1}{n}\dbinom{2n-1}{n-1}.$

The sequence

$$\binom{2n}{n}\frac{1}{4^n} = \frac{1\cdot3\cdot5\ \cdots\ (2n-3)(2n-1)}{2\cdot4\cdot6\ \cdots\ (2n-2)2n}$$

is monotone decreasing; hence,

$$\frac{1}{n}\binom{2n-1}{n-1} > \frac{1}{n+1}\binom{2n}{n}\frac{1}{4} > \frac{1}{n+2}\binom{2n+2}{n+1}\frac{1}{4^2} > \cdots .$$

Combining this result with the inequality (4.34) we find an upper bound for the remainder

$$R_nx^n + R_{n+1}x^{n+1} + R_{n+2}x^{n+2} + \cdots < \frac{1}{n}\binom{2n-2}{n-1}\frac{x^n}{1-4x} \quad \text{for } 0 < x < \frac{1}{4}.$$

The above mentioned criterion together with the estimation of the remainder provides a method to compute ρ; the computation of σ can be dealt with similarly. I have found

$$(4.35) \qquad 0.35 < \rho < 0.36, \qquad 0.30 < \sigma < 0.31.$$

Asymptotic Values of the Coefficients of Certain Power Series

75. In the preceding section the analytical behavior of the power series $q(x)$, $r(x)$, $s(x)$, $t(x)$ on the circle of convergence has been examined. An easily derived and well-known relation between the singularities and the coefficients of a power series is given below; it allows for several inferences.[1]

Lemma: *Assume that the power series*

$$f(x) = a_0 + a_1x + a_2x^2 + \cdots + a_nx^n \cdots$$

has its only singularity $x = \alpha$ on the circle of convergence. In the neighborhood of α the function has the form

$$(4.36) \qquad f(x) = \left[1 - \frac{x}{\alpha}\right]^{-s}g(x) + \left[1 - \frac{x}{\alpha}\right]^{-t}k(x),$$

where $g(x)$ and $k(x)$ are analytical functions which are regular in a neighborhood of $x = \alpha$, in particular $g(\alpha) = A \neq 0$; s and t are real constants; s is not a nonpositive integer (i.e., $s \neq 0, -1, -2, ...$); either $t < s$ or $t = 0$.

Under these conditions we have, for n increasing to infinity,

$$a_n \sim \alpha^{-n}n^{s-1}\frac{A}{\Gamma(s)}.$$

---

[1]Cf. e.g., R. Jungen, *Commentarii Math. Helvetici*, 3(1931), pp. 266-306. Theorem A, p. 269, and Theorem I, p. 275.

It is a consequence of the assumptions of the lemma that the second term of the right-hand side of (4.36) is either regular or at least "less seriously" singular than the first term on the right; the second term is, in particular, regular if $t = 0$. The proposition states that the coefficient a_n behaves asymptotically like the coefficient of x^n in the power expansion of

$$A\left(1 - \frac{x}{a}\right)^{-s}.$$

76. *Asymptotic behavior of* Q_n, R_n, S_n, T_n. The conclusion at the end of Sec. 70 can be restated as

$$(4.37) \qquad \sum_{n=0}^{\infty} Q_n x^n = q(x) = \frac{1}{1-xq(x^2)} = K\left(1 - \frac{x}{\kappa}\right)^{-1} + h(x),$$

where $h(x)$ denotes a function which is regular on the closed disk $|x| \leqslant \kappa$ and $-\kappa K$ stands for the residue of $q(x)$ at the pole $x = \kappa$. Hence,

$$K = \frac{1}{(x[xq(x^2)]')_{x=\kappa}} = \frac{1}{Q_0 + 3Q_1\kappa^3 + 5Q_2\kappa^5 + \cdots}.$$

Equation (4.37) has the form required in Sec. 75 with

$$\alpha = \kappa, \quad s = 1, \quad t = 0, \quad A = K.$$

Therefore, we conclude

$$(4.38) \qquad Q_n \sim K \kappa^{-n}.$$

The expositions in Sec. 72, specifically formula (4.26), show that the assumptions for the lemma of Sec. 75 are satisfied for the power series $r(x)$, where

$$\alpha = \rho, \quad s = -1/2, \quad t = 0, \quad A = -b.$$

Replacing the denominator by its numerical value

$$\Gamma(-1/2) = -2\Gamma(1/2) = -2\sqrt{\pi},$$

we obtain for R_n, the number of structurally isomeric alcohols $C_nH_{2n+1}OH$, the asymptotic formula

$$(4.39) \qquad R_n \sim \rho^{-n}n^{-3/2}\frac{b}{2\sqrt{\pi}}.$$

Recall the numeric estimation (4.35) of the radius ρ of $r(x)$ and the expression (4.28) of b. The asymptotic formula (4.39) seems to provide a good approximation even for only moderately large n.[1]

The derivation of the other two asymptotic formulas given by (16) in the Introduction is equally straightforward: combine the lemma of Sec. 75 with the analytic properties discussed in Sec. 73.

77. Homologous series. Consider a chemical compound containing s H-atoms which can be replaced by alkyl radicals, C_nH_{2n+1}. We call such a molecule a basic compound.[2] The group of the structure formula of the basic compound is a permutation group of degree s which interchanges the s locations of the substituted H-atoms. The cycle index of this group is a polynomial labeled $\psi(f_1, f_2, ..., f_s)$ in the notation of Sec. 25. According to Sec. 58, the number of those structurally isomeric homologues (alkyl derivatives) of the basic compound which contain n C-atoms more than the basic compound is given by the coefficient of x^n in the expansion of

$$(4.40) \qquad \psi(r(x), r(x^2), ..., r(x^s)).$$

The results of Sec. 72 imply that $x = \rho$ is the only singular point of the function (4.40) on the disk $|x| \leqslant \rho$. Expansion of (4.40) in powers of $\sqrt{x-\rho}$ beginning, according to Sec. 26, with the difference

$$(4.41) \qquad \psi(r(\rho), ..., r(\rho^s)) - \psi'_{f_1}(r(\rho), ..., r(\rho^s))b\sqrt{1 - x/\rho} + \cdots,$$

shows that $x = \rho$ is an algebraic singularity of ψ: the second term in (4.41) does not vanish because the partial derivative $\psi'_{f_1}(f_1, ..., f_s)$ has positive coefficients and all the values $r(\rho), r(\rho^2), ..., r(\rho^s)$, are positive.

The expansion (4.41) of (4.40) indicates that the lemma of Sec. 75 is applicable. We conclude that the coefficient of x^n in the MacLaurin expansion of (4.40) is asymptotically

$$\sim \rho^{-n}n^{-3/2} \frac{b}{2\sqrt{\pi}} \psi'_{f_1}(r(\rho), ..., r(\rho^s)).$$

Using the terminology of Sec. 5 of the Introduction we can state: *the number of those structurally isomeric homologues of a given basic compound which contain n carbon atoms more than the basic compound is asymptotically proportional to the number of the structurally isomeric* $C_nH_{2n+1}OH$. The formula indicates, furthermore, how the ratio depends on the index of the group of the structure formula of the basic compound.

---

[1] Cf. Pólya 5.

[2] Cayley calls it kenogram (Biggs, N. L. *Algebraic Graph Theory*, Tracts in Mathematics Series, No. 67, Cambridge University Press, 1974, p. 61).

78. *Multiply substituted paraffins.* Now we turn to the function $R(x)$ defined in (3.13). The analytic properties of $R(x)$ are discussed in Sec. 72, in particular in the expansion (4.26) and the conclusions of the section in question. With the help of Sec. 72 one easily establishes that: *The function*

$$(4.42) \qquad \frac{1}{1 - xR(x)} = \frac{1}{\rho ab} \left(1 - \frac{x}{\rho} \right)^{-1/2} + \cdots$$

is regular in the closed disk $|x| \leqslant \rho$, *with the sole exception of* $x = \rho$; *in a neighborhood of the singular point,* (4.42) *expands in increasing powers of* $\sqrt{x-\rho}$ *with* -1 *being the lowest exponent.*

Next we turn to the sum (3.18). Each of its finitely many terms is, according to the above, regular in the open disk $|x| < \rho$, with the sole singularity $x = \rho$ on the circle of convergence, and in a neighborhood of $x = \rho$ it can be expanded in increasing powers of $\sqrt{x-\rho}$. The same holds true for the sum (3.18). A term of the sum is the "more singular" the higher the exponent s in the denominator. The largest value of s occurs, according to (3.17), for

$$(4.43) \qquad p_4 = 0, \quad s = 2\ell - 3, \quad p_3 = \ell - 2.$$

There are, according to the remark at the end of Sec. 63

$$1 \cdot 3 \cdot 5 \cdots (2\ell - 5)$$

terms of the sum (3.18), which reach the highest value of s. The expansion of the sum near $x = \rho$ starts with the sum of the first terms of its summands whose parameters belong to (4.43):

$$(4.44) \qquad
\begin{aligned}
1 \cdot 3 &\cdots (2\ell - 5) \frac{[xr(x)]^{\ell-2}}{[1 - xR(x)]^{2\ell-3}} \\
&= \frac{1 \cdot 3 \cdot 5 \cdots (2\ell - 5)}{(\rho ab)^{2\ell-3}} \left(1 - \frac{x}{\rho} \right)^{-(2\ell-3)/2} + \cdots .
\end{aligned}$$

The first term on the right-hand side of (4.44) is based on (4.26) and (4.42).

Application of the lemma of Sec. 75 to (3.18) or, equivalently, to (4.44) shows: *The number of structurally isomeric* $C_n H_{2n+2-\ell} X'X'' \cdots X^{(\ell)}$ *is*

$$\sim \rho^{-n} n^{(2\ell-5)/2} \frac{1 \cdot 3 \cdot 5 \cdots (2\ell-5)(\rho a)^{\ell-2}}{(\rho ab)^{2\ell-3}} \frac{1}{\Gamma(2\ell-3)/2}$$

$$= \rho^{-n} n^{(2\ell-5)/2} \frac{\rho ab^3}{4\sqrt{\pi}} \left[\frac{2}{\rho ab^2} \right]^{\ell},$$

which shows that replacement of another H-atom multiplies the asymptotic number by a further factor, proportional to the number n of carbon atoms. This result has been announced in the Introduction. We have proved it for $\ell \geqslant 2$ but it holds also for $\ell = 1$, in which case it is equivalent to (4.39). We have not yet examined the case $l = 0$, that is, the case of the unsubstituted paraffins. We will find out whether the result holds for $l = 0$ at the end of the derivations presented in Secs. 80 - 86.

79. *Hydrocarbons* $C_nH_{2n+2-2\mu}$. We start with the case $\mu = 1$. The hydrocarbons of the form C_nH_{2n} are either homologues of alkenes or cycloparaffins, depending on whether they contain double bonds or not. The number of structurally isomeric homologues of alkenes C_nH_{2n} is by Sec. 77 asymptotically proportional to $\rho^{-n}n^{-3/2}$. The generating function P of the structurally isomeric cycloparaffins is given by (3.22). This equation implies that P has exactly one singular point on the closed disk $|x| \leqslant \rho$, namely the point $x = \rho$. Only one term, for $k = 1$, in the sum in the last line of (3.22) is singular for $x = \rho$:

$$\frac{1}{2} \log \frac{1}{1 - xR(x)} = \frac{1}{2}\log\left[\frac{1}{\rho ab}\left(1 - \frac{x}{\rho}\right)^{-1/2} + \cdots\right]$$

[cf. (4.42)]. It follows that in a neighborhood U of $x = \rho$, P is given by

$$(4.45) \qquad P = -\frac{1}{4}\log\left(1 - \frac{x}{\rho}\right) + \left(1 - \frac{x}{\rho}\right)^{1/2} g(x) + h(x),$$

where $g(x)$ and $h(x)$ are regular in U. The coefficient of x^n in the expansion of the first term on the right-hand side of (4.45) is $1/(4n\rho^n)$, which dominates $\rho^{-n}n^{-3/2}$. The lemma of Sec. 75 implies that the n-th coefficient of the last two terms on the right-hand side of (4.45) as well as the number of homologues of paraffins is asymptotically proportional to $\rho^{-n}n^{-3/2}$. Hence, *the number of structurally isomeric hydrocarbons of molecular formula* C_nH_{2n} *is asymptotically equal to*

$$\sim \rho^{-n}/4n.$$

As far as structural isomeric hydrocarbons $C_nH_{2n+2-2\mu}$ with fixed $\mu \geqslant 2$ are concerned, I leave it at a few remarks. The main term of the generating function, that is, the summand which dominates at $x = \rho$, is, apart from a constant factor, of the same type as the general term of the sum (3.18). The numbers p_2, p_4, s are tied to μ [cf. (2.1), (2.2), (2.3)] by the following relations:

$$(4.46) \qquad s - p_3 - p_4 + 1 = \mu, \quad 3p_3 + 4p_4 = 2s$$

which leads to

$$s = 3(\mu - 1) - p_4.$$

The largest value of s obtains for $p_4 = 0$. The summand in (3.18) becomes, for $p_4 = 0$,

$$\frac{[xr(x)]^{2(\mu-1)}}{[1 - xR(x)]^{3(\mu-1)}} = C\left(1 - \frac{x}{\rho}\right)^{-(3\mu-3)/2} + \cdots .$$

Invoking the lemma of Sec. 75 we can state: *the number of hydrocarbons* $C_nH_{2n+2-\mu}$ *with fixed* μ *is asymptotically proportional to* $\rho^{-n}n^{(3\mu-5)/2}$.

This asymptotic result agrees for $\mu = 1$ with the one derived in some detail for C_nH_{2n}. It holds also for $n = 0$ for which special value it is, however, harder to prove; see the following sections.

The Number of Structurally Isomeric Paraffins

80. I will prove the second, concerning ρ_n, of the four asymptotic formulas quoted in (17). It takes a middle position: the formula for κ_n is much simpler, the one for σ_n equally difficult, and the one for τ_n harder to prove than the formula for ρ_n. Cayley has looked at both ρ_n and τ_m; though ρ_n caught more attention, likely because of its importance in chemistry, than the number τ_n.

We have to show that

$$\rho_n\rho^n n^{5/2}$$

converges with $n \to \infty$ to a positive limit.
According to (2.41)

(4.47) $\rho_n\rho^n n^{5/2} = \rho'_n\rho^n n^{5/2} + \rho''\rho^n n^{5/2}.$

We deduce from (2.50) that

(4.48) $\rho'_n\rho^n n^{5/2} = \rho\left[\frac{1}{24}U_n + \frac{1}{4}V_n + \frac{1}{8}W_n + \frac{1}{3}W'_n + \frac{1}{4}W''_n\right],$

where

(4.49) $U_n = n^{5/2} \Sigma R_i\rho^i R_j\rho^j R_k\rho^k R_\ell\rho^\ell,$ $(i + j + k + \ell = n - 1),$

(4.50) $V_n = n^{5/2} \Sigma R_i\rho^i R_j\rho^j R_k\rho^{2k},$ $(i + j + 2k = n - 1),$

(4.51) $W_n = n^{5/2}\rho^{(n-1)/2} \Sigma R_i\rho^i R_j\rho^j,$ $(2i + 2j = n - 1),$

(4.52) $W'_n = n^{5/2}\rho^{(n-1)/3} \Sigma R_i\rho^i R_j\rho^j \cdot \rho^{2j-(n-1)/3},$ $(i + 3j = n - 1),$

(4.53) $W''_n = n^{5/2}\rho^{3(n-1)/4} \Sigma R_i\rho^i,$ $(4i = n - 1).$

The summations extend over sets of integers (i,j,k,ℓ), (i,j,k), (i,k), (i), respectively, which satisfy the indicated equations as well as the inequalities

$$(4.54) \qquad 0 \leqslant i < \frac{n}{2}, \quad 0 \leqslant j < \frac{n}{2}, \quad 0 \leqslant k < \frac{n}{2}, \quad 0 \leqslant \ell < \frac{n}{2}.$$

"Empty" sums are interpreted as zeros. For example, W_n'' consists of a single term or assumes the value 0 depending on whether 4 is a divisor of $n - 1$ or not.

To evaluate the quantities introduced we mainly use relation (4.39), whence we conclude that there exists a positive number C such that

$$(4.55) \qquad R_m \rho^m m^{3/2} < C \qquad \text{for} \quad m = 1, 2, 3, \ldots ,$$

and furthermore that the three positive series

$$(4.56) \qquad R_0 + R_1 \rho + \cdots + R_k \rho^k + \cdots = r(\rho) = a,$$

$$(4.57) \qquad 1 R_0 + 3 R_1 \rho^2 + \cdots + (2k + 1) R_k \rho^{2k} + \cdots ,$$

$$(4.58) \qquad 1 R_0^2 + 3 R_1^2 \rho^2 + \cdots + (2k + 1) R_k^2 \rho^{2k} + \cdots$$

converge. The inequality

$$(4.59) \qquad 0 < \rho < 1$$

will have to be taken into account.

81. Relations (2.46) and (4.39) imply

$$(4.60) \qquad \rho_n'' \rho^n n^{5/2} \leqslant R_{n/2}^2 \rho^n n^{5/2} = O(n^{-1/2}).$$

Furthermore, it is easy to confirm

$$(4.61) \qquad W_n \to 0, \quad W_n' \to 0, \quad W_n'' \to 0.$$

For W_n'', the convergence of $r(\rho)$ and $\rho < 1$ suffice; for W_n, we note that the square of the absolutely convergent series $r(\rho)$ converges too; for W_n', the inequality, resulting from (4.54) and the condition imposed on the summation (4.52),

$$3j = n - 1 - i \geqslant \frac{n - 1}{2}, \quad \text{i.e.} \quad 2j \geqslant \frac{n - 1}{3},$$

has to be added.

82. The expression V_n is harder to deal with. We split the sum (4.50) into two parts: the terms for which

$$(4.62) \qquad k \leqslant (n - 3)/6$$

make up V_n', the remainder V_n'' so that

(4.63) $V_n = V_n' + V_n''$.

The second sum

$$V'' = n^{5/2} \rho^{(n-3)/6} \Sigma \, R_i \rho^i R_j \rho^j R_k \rho^k \rho^{k-(n-3)/6}$$

contains by definition only terms for which the exponent of the last factor is positive. Since also the third power of $r(\rho)$ converges, we conclude

(4.64) $V_n'' \to 0$.

With the help of (4.55) a dominant for V_n' is found:

(4.65) $$V_n' = n^{5/2} \sum_{k=0}^{(n-3)/6} R_k \rho^{2k} \sum_{i,j} R_i \rho^i R_j \rho^j$$

$$\leqslant n^{5/2} \sum_{k=0}^{(n-3)/6} R_k \rho^{2k} C^2 \mu^{-3} L(k),$$

where μ denotes the smaller (not larger) of the two numbers i and j and $L(k)$ stands for the number of pairs i,j which, in addition to (4.54) satisfy the equation

(4.66) $i + j = n - 2k - 1$.

Because of (4.54) and (4.62) we have

(4.67) $n - 1 = i + j + 2k < \mu + \dfrac{n}{2} + \dfrac{n + 3}{3}$, $\mu > \dfrac{n}{6}$.

Rewriting (4.66) for n even and n odd as

$$\left[\frac{n}{2} - i \right] + \left[\frac{n}{2} - j \right] = 2k + 1$$

and

$$\left[\frac{n + 1}{2} - i \right] + \left[\frac{n - 1}{2} - j \right] = 2k + 1,$$

respectively, we see that

(4.68) $L(k) \leqslant 2k + 1$.

Relations (4.65), (4.67), (4.68) imply

(4.69) $V_n' < n^{-1/2} C^2 6^3 \sum_{k=0}^{(n-3)/6} (2k + 1) R_k \rho^{2k} = O(n^{-1/2})$,

because the series (4.57) converges.

83. The sum U_n, (4.49), is split into three parts,

(4.70) $U_n = U_n' + U_n'' + U_n'''$.

The last, U_n''', contains the terms of the sum in which all four summation indices, i, j, k, ℓ, are larger than $(n-3)/6$.

U_n'' consists of the terms for which the smaller two indices are both $\leqslant (n-3)/6$.

U_n' comprises all the terms in which only one of the indices i, j, k, ℓ is $\leqslant (n-3)/6$.

The sum U_n, (4.49), consists of fewer than n^3 terms. Due to (4.55), each of the four factors in U_n is of order

$$O(n^{-3/2}),$$

and thus

(4.71) $U_n''' = O(n^{5/2}n^{-12/2}n^3) = O(n^{-1/2})$.

The part of U_n'' which stems from $k = 1$ is not larger than

$$n^{5/2} \sum_{k=0}^{(n-3)/6} R_k^2 \rho^{2k} \sum_{i,j} R_i \rho^i R_j \rho^j$$

$$< n^{5/2} \sum_{k=0}^{(n-3)/6} R_k^2 \rho^{2k} C^2 \mu^{-3} L(k)$$

$$< n^{-1/2} C^2 6^3 \sum_{k=0}^{(n-3)/6} (2k + 1) R_k^2 \rho^{2k} = O(n^{-1/2}).$$

We use the same notation μ, $L(k)$ and the same estimations (4.67), (4.68) as in Sec. 82. The only difference in the estimations of V_n' in Sec. 82 and of U_n''' here consists in the use of the convergence of the series (4.57) for V_n' and of the series (4.58) for U_n''.
We note that

(4.72) $U_n'' = O(n^{-1/2})$.

84. We partition the terms of the sum $n^{-5/2}U_n'$ into four categories, depending on which of the four indices is the smallest of the quadruple. By definition of U_n' the smallest index is uniquely determined. The terms of the four categories add up to the same value. Singling out ℓ we write

(4.73) $U_n' = 4 \sum_{\ell=0}^{(n-3)/6} R_\ell \rho^\ell U_{n,\ell}$,

where

$$U_{n,\ell} = n^{5/2} \sum_{i,j,k} R_i \rho^i R_j \rho^j R_k \rho^k$$

(4.74)

$$= \sum_{i,j,k} R_i \rho^i i^{3/2} R_j \rho^j j^{3/2} R_k \rho^k k^{3/2} \left(\frac{i}{n} \frac{j}{n} \frac{k}{n} \right)^{-3/2} n^{-2}.$$

Summation extends over all triples of integers (i, j, k) for which

$$i + j + k = n - \ell - 1$$

$$\ell < i < \frac{n}{2}, \qquad \ell < j < \frac{n}{2}, \qquad \ell < k < \frac{n}{2}.$$

Equation (4.74) together with (4.39) indicates that for fixed ℓ and increasing n

$$U_{n,\ell} \sim \left(\frac{b}{2\sqrt{\pi}} \right)^3 \sum_{i,j,k} \left(\frac{i}{n} \frac{j}{n} \frac{k}{n} \right)^{-3/2} n^{-2}$$

(4.75)

$$\rightarrow \left(\frac{b}{2\sqrt{\pi}} \right)^3 \iint_D [xy(1 - x - y)]^{-3/2} dx\, dy$$

$$= \left(\frac{b}{2\sqrt{\pi}} \right)^3 I.$$

The domain D of integration is characterized by the inequalities

$$0 \leqslant x \leqslant \frac{1}{2}, \qquad 0 \leqslant y \leqslant \frac{1}{2}, \qquad 0 \leqslant 1 - x - y \leqslant \frac{1}{2},$$

or, equivalently, by

(4.76) $\qquad x \leqslant \dfrac{1}{2}, \qquad y \leqslant \dfrac{1}{2}, \qquad x + y \geqslant \dfrac{1}{2};$

D is a triangle. The integral in (4.75) is, in Riemann's definition, an improper integral. It can be approximated by a finite sum because the integrand is monotone, which implies, furthermore, that $U_{n,\ell}$ is bounded. I skip the details because considerations of this type are standard in the discussion of improper Riemannian integrals.[1]

Equations (4.73) and (4.75), in the notation of (4.56), imply

(4.77) $\qquad \lim_{n \to \infty} U'_n = 4 \left(\dfrac{b}{2\sqrt{\pi}} \right)^3 I \sum_{\ell=0}^{\infty} R_\ell \rho^\ell = 4a \left(\dfrac{b}{2\sqrt{\pi}} \right)^3 I.$

85. The area integral I of (4.75) is easy to evaluate. We have

(4.78) $\qquad I = \displaystyle\int_0^{1/2} y^{-3/2} K\, dy$

for

---

[1] Pólya and Szegö, *Problems and Theorems in Analysis*, Vol. 1, pp. 51-53, 236 (p. 53, problem 25).

$$K = \int_{1/2-y}^{1/2} \frac{dx}{x^3[(1-x-y)/x]^{3/2}} = \frac{2}{(1-y)^2} \int_{\sqrt{1-2y}}^{1/\sqrt{1-2y}} \frac{1+t^2}{t^2} \, dt$$

$$= \frac{8y}{(1-y)^2 \sqrt{1-2y}} \ ,$$

where $t^2 = (1-x-y)/x$. Replacing K in (4.78), we get

$$(4.79) \qquad I = 8 \int_0^{1/2} \frac{dy}{(1-y)^2 \sqrt{y(1-2y)}} = 16\sqrt{2} \int_0^1 \frac{x^{-1/2}(1-x)^{-1/2}}{(1+x)^2} dx$$

with the substitution $1 - 2y = x$. A special case of an integral evaluated by Abel[1] is

$$\int_0^1 \frac{x^{-1/2}(1-x)^{-1/2}}{u+x} = \pi(u+u^2)^{-1/2}.$$

Differentiating this equation first with respect to u and then setting $u = 1$, we find I of (4.79):

$$(4.80) \qquad I = \int_D\!\!\int [xy(2-x-y)]^{-3/2} dx \, dy = 12\pi.$$

86. Combining (4.47), (4.48), (4.60), (4.61), (4.63), (4.64), (4.69), (4.70), (4.71), (4.72), (4.77), and (4.80), we obtain

$$\lim_{n\to\infty} P_n \rho^n n^{5/2} = \frac{\rho}{24} \lim_{n\to\infty} U_n' = \frac{\rho}{24} \, 4a \left[\frac{b}{2\sqrt{\pi}} \right]^3 12\pi$$

or

$$(4.81) \qquad P_n \sim \rho^{-n} n^{-5/2} \cdot \frac{\rho a b^3}{4\sqrt{\pi}} \ .$$

This indicates that the result of Sec. 78 holds also for $\ell = 0$. The derivation in Sec. 78 made it plausible but did not prove it. Comparison of (4.39) and (4.81) with the chemically (or combinatorially) obvious second inequality of (11) implies

$$\rho a b^2 \geqslant 2.$$

Numerical evaluation shows that inequality holds.

The curious combinatorial number P_n, that is, the number of the structurally isomeric paraffins $C_n H_{2n+2}$, which intrigued Cayley, is asymptotically given by (4.81).

[1]N. H. Abel, *Oeuvres* (1881), Vol. 1, p. 254.

References

C. M. Blair and H. R. Henze, *Journal of the American Chemical Society* 1. *53* (1931), pp. 3042-3046; 2. *53* (1931), pp. 3077-3085; 3. *54* (1932), pp. 1098-1106; 4. *54* (1932), pp. 1538-1545; 5. *55* (1933), pp. 680-686; 6. *56* (1934), p. 157.

A. Cayley, Collected mathematical papers (Cambridge, 1889-1898) 1. Vol. 3, pp. 242-246; 2. Vol. 4, pp. 112-115; 3. Vol. 9, pp. 202-204; 4. Vol. 9, pp. 427-460; 5. Vol. 9, pp. 544-545; 6. Vol. 10, pp. 598-600; 7. Vol. 11, pp. 365-367; 8. Vol. 13, pp. 26-28.

C. Jordan 1. *J. f. die reine und angewandte Math.* **70** (1869), pp. 185-190.

D. König 1. *Theorie der endlichen und unendlichen Graphen* (Leipzig 1936).

A. C. Lunn and J. K. Senior, *Journal of Physical Chemistry* **33** (1929), pp. 1027-1079.

G. Pólya 1. *Helvetica chimica Acta* **19** (1936), pp. 22-24; 2. *Comptes Rendus, Académie des Sciences* **201** (1935), pp. 1167-1169; 3. *Ibid.* **202** (1936), pp. 1554-1556; 4. *Zeitschr. f. Kristallographie* (A) **93** (1936), pp. 415-441; 5. *Vierteljahresschrift d. Naturf. Ges. Zürich* **81** (1936), pp. 243-258.

G. Pólya and G. Szegö, *Problems and Theorems in Analysis*, Vol. 1, Springer-Verlag, New York, 1972.

THE LEGACY OF PÓLYA'S PAPER: FIFTY YEARS OF PÓLYA THEORY

by Ronald C. Read
University of Waterloo

1. Introduction: The General Idea of Pólya's Theorem

Pólya's paper, translated here for the first time, was a landmark in the history of combinatorial analysis. It presented to mathematicians a unified technique for solving a wide class of combinatorial problems -- a technique which is summarized in Pólya's main theorem, the "Hauptsatz" of Section 16 of his paper, which will here be referred to as "Pólya's Theorem". This theorem can be explained and expounded in many different ways, and at many different levels, ranging from the down-to-earth to highly abstract. It will be convenient for future reference to review the essentials of Pólya's Theorem, and to this end I offer the following, rather mundane, way of looking at the type of problem to which the theorem applies and the way that it provides a solution.

Suppose we have a set of boxes which for the moment we shall assume are all distinguishable one from the other. Next we have a "store" of objects, called figures, and we are going to put exactly one figure in each box. We are allowed to put the same figure in several different boxes. With each figure is associated a non-negative integer -- its "content". The result of all this -- the set of boxes, each containing a figure -- will be called a configuration. In more abstract terms, a configuration is a mapping from a set D (the boxes) to a set R (the figures). Each configuration will have a content, namely the sum of the contents of the figures in the boxes. The basic problem then is: given the relevant information about the figures and their contents and the number of boxes, how many different configurations are there with given content k?

One of the attractions of Pólya's paper is the elegant way that generating functions are introduced and used. In order to summarize the information about the figures and their contents, Pólya introduced the "figure generating function" (also frequently called the "figure counting series")

$$f(x) = a_0 + a_1 x + a_2 x^2 + a_3 x^3 + \cdots$$

where a_r is the number of figures having content r. The solutions to the problem just given, for all values of n, are displayed in another generating function or counting series, the "configuration generating function"

$$F(x) = A_0 + A_1 x + A_2 x^2 + A_3 x^3 + \cdots$$

where A_k is the number of configurations of content k. For the problem just given, with distinguishable boxes, it is quite easy to see by elementary arguments that

$$F(x) = [f(x)]^n$$

where n is the number of boxes. This elementary result can also be derived from Pólya's Theorem, but to do so would be using a steamhammer to crack a nut. To exhibit properly the power of the steamhammer we must turn to problems where the boxes are not all distinguishable.

Suppose we have a group G of permutations of n objects, in our case the boxes. We say that two configurations are equivalent if one can be obtained from the other by permuting the boxes by an element of G. For example, if the boxes are all identical, then two configurations which can be obtained from each other by *any* permutation of the boxes will be indistinguishable, and hence will be regarded as equivalent; to all intents and purposes the *same* configuration. In this case the relevant group G would be the symmetric group S_n of all permutations of the boxes. At the other extreme, if the boxes are all distinct, then the only allowable permutation is the identity, and G is the group consisting of the identity permutation alone.

To make use of the group G we need some way of summarizing those properties of the group that are relevant to the problem. This was provided by Pólya in the form of the "cycle index". It is well known that a permutation can be expressed as a product of disjoint cycles. A permutation will be said to have cycle-type $(j_1, j_2, \ldots, j_n)$ if, expressed in this way, it gives j_1 cycles of length 1, j_2 of length 2, and so on. With such a permutation we associate a monomial $s_1^{j_1} s_2^{j_2} \ldots$ in indeterminates s_i (which, for a reason to be seen later, are better denoted this way than by Pólya's f_i). The average of these monomials over all the elements of G is called the cycle index of G, denoted $Z(G; s_1, s_2, \ldots)$ or just $Z(G)$. Thus

$$Z(G; s_1, s_2, \ldots) = \frac{1}{|G|} \sum_{g \in G} s_1^{j_1} s_2^{j_2} \ldots .$$

By collecting like terms this can also be written

$$Z(G; s_1, s_2, ...) = \frac{1}{|G|} \sum_{(j)} A_{(j)} s_1^{j_1} s_2^{j_2} \cdots ,$$

where the (j) denotes summation over all non-negative solutions of

$$j_1 + 2j_2 + 3j_3 + \cdots + nj_n = n,$$

an obvious necessary condition on the j_i, and $A_{(j)}$ denotes the number of elements of G of cycle type $(j_1, j_2, ...)$.

We can now display Pólya's Theorem in its one-variable form as follows.

Theorem. *The configuration generating function is obtained by substituting the figure generating function in the cycle index, by which is meant replacing every occurrence of s_i in the cycle index by $f(x^i)$. Thus*

$$F(x) = Z(G; f(x), f(x^2), f(x^3), ...).$$

More generally the content of a figure will be a vector of non-negative integers. Pólya frequently used vectors of dimension 3. In that case the generating functions will be functions of three variables, and the statement of Pólya's Theorem then gives

$$F(x,y,z) = Z(G; f(x,y,z), f(x^2,y^2,z^2), ...).$$

To illustrate the foregoing points, here is a typical problem.

Suppose we wish to enumerate necklaces of six beads, each of which is either red or green. There are thus two figures, and six boxes (locations) in which to put them. Since we may rotate the necklace or turn it over, the appropriate group is the dihedral group D_6. Let "content" denote "number of green beads". The figure generating function is then $1 + x$. Since

$$Z(D_6) = \tfrac{1}{12}(s_1^6 + 2s_3^2 + 4s_2^3 + 3s_1^2 s_2^2 + 2s_6),$$

we see that the configuration generating function is

$$Z(D_6; 1+x) = \tfrac{1}{12}[(1+x)^6 + 2(1+x^3)^2 + 4(1+x^2)^3$$

$$+ 3(1+x)^2(1+x^2)^2 + 2(1+x^6)].$$

This seemingly almost frivolous example relates directly to a problem of chemistry. The benzene molecule consists of six carbon atoms in a ring, to each of which is attached a hydrogen atom (see Figure 1). Derivatives of benzene are formed by replacing the hydrogen atoms by other atoms or groups of atoms. If we allow just one kind of substituent, say chlorine, then the problem of determining the number of substituted benzenes is essentially the necklace problem just mentioned. If a wider set of substituents is allowed, then the simple figure generating function $1 + x$ must be replaced by the

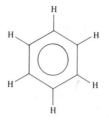

Figure 1. The benzene molecule.

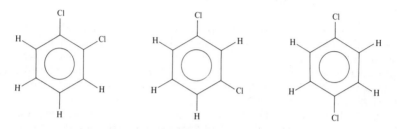

Figure 2. The three dichlorobenzenes.

appropriate figure generating function for that set. In this way Pólya's Theorem can be used to enumerate chemical "isomers", that is, molecules which have the same numbers of atoms of the various kinds, but which, because the atoms are differently bonded to each other, are chemically distinct. Figure 2 shows the three isomers resulting from the three ways of substituting chlorine atoms for two of the hydrogen atoms in benzene. This corresponds to the coefficient of x^2 in the generating function for the necklace problem above.

A variation of the basic problem with indistinguishable boxes occurs when the mapping from boxes to figures is required to be one-to-one, that is, when we are not allowed to put the same figure in two different boxes. In that case Pólya showed that the substitution of the figure generating function must be made, not in the cycle index $Z(S_n)$, but in the expression

$$Z(A_n) - Z(S_n)$$

where A_n is the alternating group. This expression is the same as $Z(S_n)$ except that those terms corresponding to odd permutations have a negative sign.

We note here an important concept introduced by Pólya, that of the "corona" or "wreath product" of two permutation groups. If M is an $m \times n$ matrix, and G and H are permutation groups of degree m and n, consider all permutations of the mn elements of M obtained as follows:

(a) for each row of M permute its elements by some element of H;
(b) permute the rows of M among themselves by an element of G.

The resulting group was denoted $G[H]$ by Pólya, who showed that its cycle index can be obtained by "substituting" the cycle index of H into that of G. By this is meant replacing every occurrence of s_i in $Z(G)$ by $Z(H; s_i, s_{2i}, s_{3i}, \ldots)$. This kind of group is of common occurrence in combinatorial problems.

2. Before Pólya's Paper

It would be quite wrong to suppose that, in 1937, Pólya's paper and its results leapt fully armed from Pólya's brain, like Athene from the head of Zeus. Quite the contrary. Pólya's 1937 paper was the culmination and collation of work that had been under study for some time previously, and much of this work had already appeared in print.

An earlier paper [PolG35] contained the essence of Pólya's Theorem in its three-variable form, together with an informal proof of it. This paper also entered into the field of isomer enumeration by presenting a generating function for the substituted benzenes.

A paper in the same journal [PolG36b] elaborated on isomer enumeration and the corresponding asymptotic results. Here the functional equations for the generating functions for four kinds of rooted trees were presented without proof. They were, in a slightly different notation, formulae (8), (4), and (7) in the introduction to Pólya's main paper, and one form of the functional equation for the generating function for rooted trees. From these results a number of asymptotic formulae were derived. These results were all incorporated into the main paper.

Rather similar was the paper [PolG36a] which also derives asymptotic formulae for the number of several kinds of chemical compounds, for example the alcohols and benzene and naphthalene derivatives. Unlike the paper previously mentioned, this one gives proofs of the recursion formulae from which the asymptotic results are derived. A third paper on this topic [PolG36] covers the same sort of ground but ranges more broadly over the chemical compounds. Derivatives of anthracene, pyrene, phenanthrene, and thiophene are considered as well as primary, secondary, and tertiary alcohols, esters, and ketones. In this paper Pólya addresses the question of enumerating stereoisomers -- a topic to which we shall return later.

3. Burnside's Lemma

Pólya's Theorem depends on a group theoretical result which now tends to be called Burnside's Lemma, by virtue of its appearance in Burnside's book [BurW11]. This is somewhat of a misnomer since the lemma was certainly known to Frobenius, and, in its essence, can be traced back to Cauchy. For historical information on this point see [NeuP79], [WriE81], and especially the amusing footnote to page 101 of [PalE85]. However, attempts to call the lemma something else, such as the Cauchy-Frobenius theorem, or otherwise correct the mis-attribution, have not met with much success, and at the risk of seeming unjust I shall use the commonly accepted ascription. It is not uncommon in mathematics for a theorem to be named, not after the discoverer, but after someone who has popularized it or in whose works it happens to be conveniently accessible, or even after some-one with no claim whatever to such eponymous immortality! In any case, it behooves us not to be too critical, since, as we shall see in Section 8, it can be argued that "Pólya's Theorem" is also a misnomer!

Burnside's Lemma in its simplest form states that, under condi-tions similar to those of the general Pólya problem given above, the number of inequivalent configurations can be obtained as follows. For each element of G count the number of configurations that are invariant under G, and take the average of these numbers. In symbols, the number of inequivalent configurations is

$$\frac{1}{|G|} \sum_{g \in G} N(g),$$

where $N(g)$ is the number of configurations invariant under g. This lemma has been likened to the country yokel's method for counting cows, namely "count the legs and divide by four". In adding up the various counts (as given by the summation sign), each equivalence class of configurations is counted many times; the force of the lemma resides in the fact that the equivalence classes end up all being counted the same number of times, namely $|G|$.

Looking at this result from a more abstract, group theoretical, point of view, we see that the equivalence classes of configurations correspond to orbits of the group G. Hence Burnside's Lemma effectively counts the number of orbits. Let us now suppose that each configuration c has a "weight" $w(c)$, and that equivalent con-figurations have the same weight, so that the weight function is con-stant over an orbit of G. We can then ask for the sum of the weights of the orbits. This is given by an extended (weighted) form of Burnside's Lemma as

$$\sum_{\text{all orbits}} W(C) = \frac{1}{|G|} \sum w(c),$$

where C denotes a general orbit, and the summation on the right-hand side is over all pairs (c,g) such that c is invariant under g.

Clearly, if we put $w(c) = 1$ for every configuration we recover the original, unweighted, form of Burnside's Lemma.

To obtain Pólya's Theorem from this we take the "weight" of a figure to be x^k, where k is its content. (We should note here that the term "weight" is regarded by some authors as synonymous with "content"; however, it is preferable to use them for the two different concepts just given.) Suppose that a configuration is invariant under

a permutation of cycle-type $s_1^{j_1} s_2^{j_2}, \dots$. Then for each cycle of length r we must place the same figure, of content k say, in each of the r boxes. These boxes then behave like a single box for which the figure has content kr. This is what lies behind the substituting of $f(x^r)$ in place of s_r in the statement of Pólya's Theorem, and is the main step in its derivation.

4. Early Expositions of Pólya's Theorem

It used to be the case that almost any author who wrote a paper on some application of Pólya's Theorem deemed it necessary to preface the application with an account, perhaps brief, perhaps comprehensive, of the theorem. Thus the number of expositions of the theorem in the early literature was almost equal to the number of publications. In this section we shall look only at some early expositions that were given of the theorem for its own sake, for the purpose of instruction, rather than as a means to an end.

Way back in the early 1950s books on graph theory or on the kind of combinatorics with which Pólya's Theorem is concerned were virtually nonexistent. In graph theory the only readily available book was that of König [KonD36], and the only comprehensive source of information about Pólya theory was Pólya's paper itself. Thus the theory remained largely inaccessible to those mathematicians who did not read German. This situation improved in 1958 when Riordan's book on combinatorial analysis [RioJ58] appeared, containing among much other material a brief account of Pólya's Theorem. The same year saw the publication of Berge's *Theorie des Graphes* [BerC58] which carried in an appendix an excellent account of Pólya's Theorem by J. Riguet. Thus information on the theorem became available also to readers of English and French. (Berge's book was translated into English four years later.) Nevertheless, many mathematicians owe their first acquaintance with Pólya theory to the chapter by N. de Bruijn on "Pólya's Theory of Counting" in the book *Applied Combinatorial Mathematics* edited by Beckenbach [deBN64], though, of course, the brief *ad hoc* accounts in various papers were also available and useful.

The mid-1960s saw the beginning of the explosion of publications in graph theory and combinatorics, which shows no signs of abating, and from the mid-1960s onwards, expositions of Pólya theory from varying points of view came thick and fast.

In 1967 Harary presented a proof of Pólya's Theorem in Chapter 4 of his book [HarF67], and in the following chapter gave some applications. Four years later in Chapter 15 of "Graph Theory" [HarF71] he again gave an exposition of Pólya's Theorem and some developments of it.

Comtet's two-volume work on combinatorics [ComL70] appeared in 1970. It contained an account and proof of Pólya's Theorem together with all the necessary preliminaries -- definitions of cycle-index, Burnside's Lemma, and so on. Comtet illustrated Pólya's Theorem by a single example, the coloring of the faces of a cube.

Also in that year appeared a set of notes by Garsia [GarA71] which has been quoted several times and is evidently well esteemed; unfortunately I have not seen a copy of it.

The next milestone was the 1973 monograph on graphical enumeration by Harary and Palmer [HarF73] which covered in great detail those combinatorial techniques, including the use of Pólya's Theorem and its generalizations, appropriate to the enumeration of many kinds of graphs and graph-like objects. Among much else, the authors gave an authoritative account of their power group enumeration theorem and its derivation from Pólya's Theorem via the power group of two groups, together with a wealth of illustrative examples. Graphical enumeration in general, and this book in particular, are discussed elsewhere in this article.

Meanwhile the "Introduction to Combinatorial Mathematics" by C. L. Liu had appeared [LiuC68]. Chapter 5 of this book gives a full account of Pólya's theory of counting, including a proof of the theorem using Burnside's Lemma, and several examples -- a necklace problem, painting the faces of a cube, and a simple example of chemical enumeration. It also discusses de Bruijn's generalization of Pólya's Theorem together with appropriate examples.

During the 1950s and 60s the study of Pólya theory was probably confined mostly to established mathematicians, graduate students, and perhaps a few undergraduates. By the 1970s, interest in the theory had percolated downwards sufficiently that the time was ripe for the appearance of expositions more suited to undergraduate courses. Two such can be mentioned. A second book by Liu, "Topics in Com- binatorial Mathematics" [LiuC72], the outcome of a seminar for college teachers held in 1972, expounds (in Chapter IX) both the Pólya and de Bruijn theorems together with several easily comprehended examples. Neither theorem is proved. Alan Tucker, in his account of Pólya enumeration [TucA74] unashamedly omits any proofs, pointing out the well-known (but not often admitted) fact that students do not need to know the proof of a theorem in order to use it. Instead, in his "cookbook" approach to the subject, he gently leads the reader through a simple necklace problem (four beads, two colors), discussing the questions of symmetry until the reader is ready to appreciate Burnside's Lemma. This leads on to the definition of cycle index, and hence by easy stages to the statement of Pólya's Theorem.

By this time Pólya's Theorem had become a familiar combinatorial tool, and it was no longer necessary to explain it whenever it was used. Despite that, expositions of the theorem have continued to proliferate, to the extent that it would be futile to attempt to trace them any further. I take space, however, to mention the unusual exposition by Merris [MerR81], who analyzes in detail the 4-bead 3-color necklace problem, and interprets it in terms of symmetry classes of tensors -- an interpretation that he has used to good effect elsewhere (see [MerR80, 80a]).

5. Enumeration of Trees

A tree is a connected graph having no circuits. Trees have been prominent in graph-theoretical studies because in addition to being related to many structures that occur frequently in real life, they have the attractive property of being much easier to handle than graphs in general. Problems that are quite intractable for graphs often turn out to be quite manageable or even easy for trees. Thus although trees are merely a special kind of graph, it will be convenient to consider their enumeration separately from that of graphs in general.

A considerable part of Pólya's paper is concerned with trees and their enumeration. Indeed, in his introduction, Pólya states that his work is a continuation of that done by Cayley on this kind of enumeration.

In chemistry, the formula of an acyclic compound with no multiple bonds corresponds in an obvious way to a tree having different kinds of vertices (corresponding to the different atoms). If double or triple bonds are present, they are best regarded as different kinds of edges, and the formula is still represented as a tree. Figure 3 shows a graphical tree, and the formula of a chemical compound having its structure.

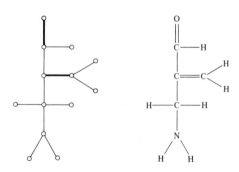

Figure 3.

A simplification in the graphical interpretation of acyclic chemical compounds is possible in the case of saturated acyclic hydrocarbons, once known as paraffins but now more usually called "alkanes". These have the general formula C_nH_{2n+2}, indicating the presence in each molecule of n carbon atoms and $2n + 2$ hydrogen atoms. If we assume that a carbon atom always has valency (or degree) 4, then we need take no account of the hydrogen atoms but look only at the structure formed by the carbon atoms; for if this structure is known, the positions for the hydrogen atoms are determined uniquely. Figure 4 shows the structural formula of an alkane (actually isobutane) and its "skeleton" of carbon atoms, which is a tree. Pólya called such a tree a "C-tree".

It was largely this chemical interpretation which led Cayley to enumerate various kinds of trees. He gave (without much of a proof) the formula n^{n-2} for the number of trees on n *labelled* vertices [CayA89], and the equation

$$(5.1) \qquad T(x) = \sum_{n=1}^{\infty} t_n x^n = x(1 - x)^{-t_1}(1 - x^2)^{-t_2}(1 - x^3)^{-t_3} \ldots$$

for the number t_n of rooted unlabelled trees on n vertices [CayA57, 59]. A tree is rooted if one of its vertices, called the root, is distinguished from the others. Cayley was also able to enumerate trees of interest in chemistry, such as rooted trees with maximum degree four. These correspond, in the manner described above, to saturated hydrocarbons in which one carbon atom is distinct from the rest. This distinction can arise, for example, if a hydrogen atom attached to that carbon has been replaced (substituted) by some other atom or group. Thus if one hydrogen is substituted by a chlorine atom, we get an alkyl chloride with the formula $C_nH_{2n+1}Cl$; if the substitution is by a hydroxyl group (OH), we get the alcohols with general formula $C_nH_{2n+1}OH$, and so on. Cayley's pioneering work in this field was embodied in a number of papers. The reader can consult [CayA74, 75, 77, 81], or the expository articles [RouD75, 77].

The use of Pólya's Theorem in the enumeration of rooted trees is amply described in Pólya's paper and needs little comment here. We shall note an important point in connection with the enumeration of alkyl radicals. A radical is a portion of a molecule that is regarded as a unit; that is, it will be treated much the same as if it were a

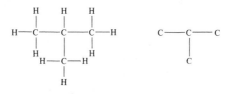

Figure 4. An alkane and its C-tree.

Figure 5. General form of an alkyl radical.

single atom, and will have a "free bond" by which it is attached to
the rest of the molecule. The terms "side-chain", "substituent", and
"ligand" are roughly synonymous with "radical". An alkyl radical is
one of the form C_nH_{2n+1}, having one free bond. Such a radical has
the form shown in Figure 5, where X, Y, Z are themselves alkyl radicals (or, as a special case, hydrogen atoms). Here we have three
boxes, and the figure generating function -- call it $s(x)$ -- is the
desired generating function for alkyl radicals -- the same function
we wish to find. But what is the group? If we assume it is S_3 then
we are allowing any permutation of X, Y, and Z, including a reflection which interchanges two of these radicals while leaving the other
fixed. This, however, is not realistic, since the radicals exist in
three-dimensional space. If X, Y, and Z are all different there are
two possible forms for the radical, and they are mirror images of
each other. Thus the more appropriate group is the alternating
group A_3. Hence the configuration generating function for the three
boxes is $Z(A_3; s(x))$. Taking into account the extra vertex to which
X, Y, and Z are attached we see that

$$(5.2) \qquad s(x) = 1 + xZ(A_3; s(x)) = 1 + \frac{x}{3}[s^3(x) + 2s(x^3)],$$

which is Pólya's equation (2.14). The 1 on the right-hand side must
be added to ensure that the constant term, corresponding to a single
hydrogen atom, is present on the left-hand side.

This illustrates an important distinction in chemical enumeration;
that between the enumeration of "structural" isomers, in which only
the connections between the atoms are considered, and that of
stereoisomers, in which the situation of a molecule in space is
important, so that as above we can have right- and left-hand forms
of a molecule. This distinction will occur, for example, when a
carbon atom is bonded to four distinct substituents (it can occur in
many other ways). Such a carbon atom is said to be asymmetrical.

Pólya's main results on tree enumeration are summarized at the
beginning of Section IV of his paper. His equation (4.8) gives the
functional equation for the generating function of rooted labelled
trees, from which Cayley's result, n^{n-2}, referred to above, follows
immediately; his (4.9) is a heavily disguised form of Cayley's
equation for the number of (unlabelled) rooted trees given here as
(5.1); while (4.10) gives the generating function for what are now
known as planted plane trees.

In the realm of chemical enumeration we note Pólya's equation (4.4) which gives the generating function for stereoisomers of the alkyl radicals, or equivalently, alcohols -- that is, equation (5.2) of this article. His equation (4.3) gives the corresponding result for the structural isomers of these compounds. His equations (4.2) and (4.5) correspond, respectively, to the cases of alcohols without any asymmetric carbon atoms and the number of embeddings in the plane of structural formulae for alcohols in general. The latter problem is not chemically very significant.

In addition to these main results some others can be found in Pólya's paper, notably the enumeration of doubly and multiply substituted alkanes.

Both Cayley and Pólya were able to enumerate unrooted trees and C-trees, but the methods they used were somewhat involved. A significant improvement in the enumeration of these trees, also known as "free" trees, was made by Otter [OttR48]. Otter's method depends on the concept of a dissimilarity characteristic, and deserves a brief description.

Let u, v be two vertices of a tree. We say they are similar if there is an automorphism of the tree which maps u onto v. This relation of similarity is an equivalence relation and partitions the p vertices of the tree into equivalence classes. Let p^* be the number of equivalence classes. Similarly we say that two edges of the tree are similar if there is an automorphism which maps one onto the other. Let q^* be the number of equivalence classes of edges under this relation. A symmetric edge in a tree is an edge, uv say, such that there is an automorphism of the tree which interchanges u and v. Let s be the number of symmetric edges in a tree; it is easy to see that s can only be 0 or 1. We then have the following theorem.

Theorem (*Dissimilarity Characteristic Theorem*). *For any tree*
$$p^* - q^* + s = 1.$$

For a proof see [HarF73, page 56].

The enumeration result that we want is obtained by summing the equation of the theorem over all *unrooted* trees with a given number p of vertices. Thus we have

(5.3) $\Sigma p^* - \Sigma q^* + \Sigma s = \Sigma 1$.

Now p^*, the number of dissimilar vertices, is precisely the number of distinct ways of rooting the tree in question; hence Σp^* will be the total number of rooted trees, which is known, since it is given by the generating function $T(x)$ defined in Cayley's result (5.1). Similarly Σq^* will be the number of trees on p vertices rooted at an edge, that is, trees in which one edge has been distinguished. This we shall determine in a moment. Σs will be the number of trees on p vertices having a symmetric edge, and the right-hand side is precisely the number of unrooted trees on p vertices.

To find the number of trees rooted at an edge we have merely to take the distinguished edge and add a rooted tree at each end. This is a Pólya-type problem with two interchangeable boxes, and figure generating function $T(x)$. Pólya's Theorem thus gives

$$\tfrac{1}{2}[T^2(x) + T(x^2)]$$

as the configuration generating function. Now such a tree will have a symmetric edge if and only if the boxes receive the same rooted tree. Hence Σs gives rise to the generating function $T(x^2)$, and thus from (5.3) we derive

$$T(x) - \tfrac{1}{2}[T^2(x) - T(x^2)]$$

as the generating function for unrooted trees. This is Otter's formula.

Well before the publication of Pólya's paper, Lunn and Senior [LunA29], Blair and Henze and Coffman [BlaC31,31a,32,32a,33,34] [CofD33], and Perry [PerD32] had extended Cayley's results to the enumeration of many other "series" of organic compounds. A summary of these results can be found in [ReaR76]. These new results are essentially just variations on Cayley's original theme, and with the advent of Pólya's theory their derivation became a matter of routine. For example, the enumeration of alkyl derivatives of acetylene, with general formula

$$X - C \equiv C - Y,$$

(where X, Y are alkyl radicals) is seen at once to be a Pólya-type problem with two interchangeable boxes and with the figure generating function $s(x)$ for the alkyl radicals. In three early papers [HilT43,43a,43b] T. L. Hill applied Pólya's Theorem to this same kind of chemical enumeration, paying special attention to various kinds of isomerism. Another early paper of a similar kind is [TayW43]. These are among the very few papers about Pólya theory to appear in the period between the publication of Pólya's paper and the early 1950s when the theorem began to be widely used.

Pólya's Theorem clearly showed the way to the general enumeration of *all* acyclic hydrocarbons, irrespective of how many double or triple bonds they might have; but it was to be 35 years before this enumeration was carried out. In two papers [ReaR72,76] I obtained the solution to this general problem in both the structural isomer and stereoisomer cases, as generating functions in three variables. Of these variables, x marks the number of carbon atoms, y the number of double bonds, and z the number of triple bonds. The de- rivation of these generating functions was Pólya theory all the way -- a succession of applications of Pólya's Theorem with occasional use of Otter's result. The derivation was really rather tedious, but the generating functions, once obtained, can be used to compute the numbers of hydrocarbons with various required characteristics. For example, the generating function for the alkyl derivatives of acetylene mentioned above can be obtained immediately from the generating function for general acyclic hydrocarbons by putting $y = 0$ and finding the coefficient of z.

A question which chemical enumerators should not ignore is that of the extent to which their results are realistic in the physical world.   Thus in [BlaC31a] it is stated that the number of alkanes (paraffins) with 40 carbon atoms is 62,491,178,805,831. Can we really be sure that all these compounds can exist; or could it be that factors not catered for in the enumeration render some of them chemically infeasible?   In this connection we should note the paper [KleD81], in which it is shown that because of such factors the chemical tree enumerations by Pólya and others give numbers that are consistently higher than the number of compounds that are in fact chemically possible.   This does not detract from the mathematical value of these results; it merely shows that care is needed in relating them to problems of real life.

6. Generalizations and Extensions of Pólya's Theorem

We have already noted that Pólya's Theorem can be stated in the context of mappings from one set to another. Let D and R be two sets, and G a group of permutations of the elements of D, and consider the set of all mappings $f: D \rightarrow R$. If D is a set of boxes and R the set of figures, then each such mapping corresponds to a configuration.

Under a permutation g of G a box $x \in D$ becomes the box $g(x)$ which maps into $f(g(x))$. Two mappings f_1 and f_2 are equivalent if there is an element g of G such that $f_1(x) = f_2(g(x))$ for every element $x \in D$; in other words, $f_1 = f_2(g)$.

An early direct generalization of Pólya's Theorem was that of de Bruijn [deBN59], which introduces the possibility of permuting the figures as well as the boxes. Thus we have a second permutation group H, acting on the set R of figures, and we allow for equivalence under the action of the group H also. That is to say, we regard two mappings f_1 and f_2 as equivalent if there is a $g \in G$ and $h \in H$ such that

$$f_1(x) = hf_2(g(x)) \quad \text{for all} \quad x.$$

As a very simple example of the application of de Bruijn's theorem, consider necklaces of six beads, of two colors, red and green. This is a straight Pólya-type problem as we have seen. If we agree to regard two necklaces as equivalent if one can be obtained from the other not only by rotating or turning the necklace, but also by interchanging the colors of the beads, then we have a problem to which de Bruijn's theorem can be applied. The group G is D_6, as with the Pólya problem, but now we have the group $H = S_2$, which interchanges two elements (figures) in R. We shall return to this problem and give its solution later.

The original formulation of de Bruijn's theorem was for a quite general problem of this type, with a broad definition of the "weight" of a mapping. We assume that R is the union of a finite number of pairwise disjoint sets R_i ($i = 1, ..., k$), and that H is a direct product of groups H_i, where H_i acts on R_i. For each R_i there is a weight function $\psi_i(n)$, where n is the number of elements of D that are mapped onto an element of R_i. The weight of a mapping is then the product of the numbers $\psi_i(n)$ summed over all elements of R. We can now state the theorem.

de Bruijn's Theorem. *For the problem just defined, the sum of the weights over the equivalence classes of functions is*

$$Z\left[A; \frac{\partial}{\partial z_1}, \frac{\partial}{\partial z_2}, ...\right]_0 \sum_{i=1}^{k} Z(B_i; \eta_{i1}, \eta_{i2}, ...)$$

where

$$\eta_{it} = \sum_{n=0}^{\infty} Z(S_n; tz_t, 2tz_{2t}, 3tz_{3t}, ...)[\psi_i(n)]^t$$

$$(i = 1, 2, ..., k; \quad t = 1, 2, ...).$$

The zero subscript indicates that the variables $z_1, z_2, ...$ are to be put equal to zero after the differentiations have been performed.

If the weight of a figure is taken to be the Pólya-type weight, that is, x^k where k is the content of the figure, then this result can be manipulated into a more convenient form. Examples of this, with some applications, can be found in [ReaR68a]. In particular, if H is the identity group, then de Bruijn's theorem reduces to Pólya's theorem. A fuller, embellished, version of de Bruijn's theorem was presented in [deBN71a], where de Bruijn christens it the "Monster theorem". Other papers by de Bruijn [deBN63,67,69,71,72] elaborate on the theorem and its applications.

It was around this time that Gian-Carlo Rota was developing his elegant theory of Möbius Inversion, a theory with far-reaching consequences in many branches of combinatorial mathematics. The basic paper on this topic is [RotG64]. Möbius inversion can be used to derive a proof of Pólya's Theorem which is not only different from the usual proof via Burnside's Lemma, but also has potential for generalization and application in new directions.

A theorem which, at first sight, does not seem to be very closely related to Pólya's Theorem, but which in fact has much affinity with it, is the superposition theorem that appeared in my doctoral thesis [ReaR58] and later in [ReaR59,60]. The general problem to which it applies is the following. Consider an ordered set of k permutation groups of degree n, say $G_1, G_2, ..., G_k$, and the set of all k-ads

$(a_1, a_2, ..., a_k)$ where $a_i \in G_i$. We say that two k-ads $(a_1, a_2, ..., a_k)$ and $(b_1, b_2, ..., b_k)$ are similar if and only if there exists a permutation $x \in S_n$ and $g_i \in G_i$ such that

$$b_i = x \, a_i \, g_i \quad \text{for} \quad i = 1, ..., k.$$

The superposition theorem then gives a method of determining the number of equivalence classes under this relation of similarity.

This theorem has a direct interpretation in terms of graphs. Let Γ_i ($i = 1, ..., k$) be k graphs, each on the same number, p, of vertices, and let G_i be the automorphism group of Γ_i. Consider the graphs that can be obtained by superposing these k graphs on the same set of vertices. In performing this superposition we must distinguish between edges of different graphs, which we can do, for example, by supposing that each graph has edges of a distinct color. It can be shown that the isomorphism classes of the superimposed graphs are precisely the similarity classes defined above for the set of graphs $G_1, G_2, ..., G_k$. Thus the superposition theorem gives the number of distinct (i.e. nonisomorphic) superposed graphs.

The theorem is as follows. Let the cycle index of G_i be

$$Z(G_i) = \frac{1}{|G_i|} \sum_{(j)} A^{(i)}_{(j)} s_1^{j_1} s_2^{j_2} \cdots s_p^{j_p},$$

where $A^{(i)}_{(j)}$ is the number of permutations of cycle type $(j) = (j_1, j_2, ..., j_p)$ in Γ_i. Then the required number is

$$\prod \frac{1}{|G_i|} \sum_{(j)} A^{(1)}_{(j)} A^{(2)}_{(j)} \cdots A^{(k)}_{(j)} (1^{j_1} j_1! \, 2^{j_2} j_2! \cdots p^{j_p} j_p!)^{k-1}.$$

This result can be made more transparent by introducing an operation, denoted $*$, between two cycle indexes. If

$$A = \sum_{(j)} a_{(j)} s_1^{j_1} s_2^{j_2} \cdots s_p^{j_p}$$

and

$$B = \sum_{(j)} b_{(j)} s_1^{j_1} s_2^{j_2} \cdots s_p^{j_p},$$

we define

$$A*B = \sum_{(j)} a_{(j)} b_{(j)} \, \pi(i^{j_i} j_i!) s_1^{j_1} s_2^{j_2} \cdots s_p^{j_p},$$

that is, for each monomial $s_1^{j_1} ... s_p^{j_p}$, the coefficient in $A*B$ is the product of the coefficients of that monomial in A and B, multiplied by $1^{j_1} j_1! \cdots p^{j_p} j_p!$. It turns out that if A and B are the cycle indexes of two graphs, then $A*B$ is the sum of the cycle indexes of the automorphism groups of the distinct superposed graphs. If C is an expression in $s_1, s_2, ..., s_p$, denote by $N(C)$ the result of setting $s_1 = s_2 = $

$\dots = s_p = 1$.   Thus $N(C)$ is the sum of the coefficients in the expression C. For a cycle index A we have, by definition, $N(A) = 1$. From what has been stated above, it follows that $N(A*B)$ will be the num- ber of superpositions of the two graphs.   Moreover the "product" $A*B$ can be extended, by associativity, to any number of cycle indexes, and will again be the cycle index sum for the superpositions of many graphs. Hence if A_i denotes, for brevity, the cycle index of the automorphism group of Γ_i -- our previous $Z(G_i)$ -- then the number of superposed graphs is given by

$$N(A_1 * A_2 * \dots * A_k).$$

By an abuse of notation, using the graph symbol to denote the cycle index, this can be conveniently written as

$$N(G_1 * G_2 * \dots * G_k).$$

Thus if $k = 3$, and $\Gamma_1 = \Gamma_2 = \Gamma_3 = $ the circuit on four vertices, we see that the number of distinct ways of superposing three such cir- cuits (of three different colors) is given by

$$N(D_4 * D_4 * D_4).$$

Since

$$Z(D_4) = \tfrac{1}{8} [s_1^4 + 2s_4 + 3s_2^2 + 2s_1^2 s_2]$$

we have

$$N(D_4 * D_4 * D_4) = \frac{1}{8^3}[(4!)^2 + 9.(4)^2 + 27.(2^2.2!)^2 + 8.(2.2)^2] = 5.$$

The distinct superpositions are shown in Figure 6 together with their automorphism groups, from which we can verify the assertions made above about the sum of their cycle indexes.   For

$$D_4 * D_4 * D_4 = \tfrac{1}{8} [9s_1^4 + 2s_4 + 27s_2^2 + 2s_1^2 s_2]$$

$$= \tfrac{1}{8} [s_1^4 + 2s_4 + s_2^2 + 2s_1^2 s_2] + 4.\tfrac{1}{4} [s_2^4 + 3s_2^2],$$

as required.

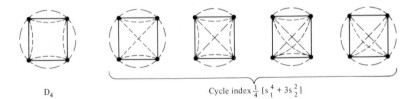

D_4 Cycle index $\tfrac{1}{4}$ $[s_1^4 + 3s_2^2]$

Figure 6.

We now look at one of the connections between the superposition theorem and Pólya's Theorem. This is that the superposition theorem can be used to find individual coefficients in a configuration generating function obtained from Pólya's Theorem, without deriving the whole series. Take the case of a necklace of six beads, two of which are red and four green. With x marking red beads and y marking green beads, the figure generating function is $x + y$, the group is D_6 and Pólya's Theorem gives $Z(D_6; x+y)$ as the configuration generating function. In this we want to find the coefficient of x^2y^4, which can easily be seen to be 3.

Consider now the two graphs in Figure 7. In (b) the two components of the graph correspond to the red beads and the four green beads. It is easy to verify that the distinct superpositions of these two graphs correspond in a one-to-one way with the required necklaces. Hence the number of necklaces is

$$N(D_6 * S_2 \times S_4)$$

since the direct product $S_2 \times S_4$ is the group of automorphisms of (b). Since

$$Z(D_6) = \tfrac{1}{12}(s_1^6 + 2s_6 + 2s_3^2 + 4s_2^3 + 3s_1^2s_2^2)$$

and

$$Z(S_2 \times S_4) = \tfrac{1}{48}(s_1^6 + 7s_1^4s_2 + 9s_1^2s_2^2 + 8s_1^3s_3 + 6s_1^2s_3$$
$$+ 3s_2^2 + 8s_1s_2s_3 + 6s_2s_4),$$

we have

$$N(D_6 * S_2 \times S_4) = \tfrac{1}{12 \cdot 48} (6! + 4.3.3!.2^3 + 3.9.2.2^2.2] = 3.$$

It goes without saying that neither Pólya's Theorem nor the superposition theorem is really needed to solve this simple problem -- a

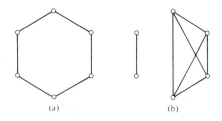

(a)   (b)

Figure 7.

moment's thought is all that is required to obtain the answer. It serves, however, to illustrate the connection between the two theorems.

An elegant general theorem encompassing the essence of Pólya's and de Bruijn's theorems was presented in 1966 by Harary and Palmer [HarF65,66]; a more definitive presentation is given in [HarF73]. It starts, as does de Bruijn's Theorem, with two sets D, R, and considers the set R^D of mappings $f: D \to R$. There is a permutation group G acting on D and another, H, acting on R, and equivalence between functions is defined as with de Bruijn's Theorem. Here, however, the functions play a role similar to that of the "boxes" in Pólya's Theorem, and the basic question is that of the nature of the group of permutations of the functions that is induced by the actions of the two groups G and H. Since this group acts on the power set R^D, it is called the power group, and denoted H^G. Harary and Palmer give a method for computing the cycle index of this group, and from this, by what is essentially an application of Polya's Theorem, they obtain their theorem, called the power group enumeration theorem (PGET).

The power group enumeration theorem has been put to a great number of uses, especially in graph theoretical and chemical enumeration. One simple example will illustrate the value of the theorem.

Let us consider again, and now solve, the necklace problem that was mentioned in the discussion of De Bruijn's Theorem, namely, to enumerate necklaces of six beads of two colors, red and green, where the two colors can be interchanged. We shall ask only for the total number, and can therefore use a simpler (unweighted) form of the power group enumeration theorem which gives as the required number the expression

$$(6.1) \qquad \frac{1}{|H|} \sum_{h \in H} Z(G; c_1(h), \ldots, c_m(h))$$

where

$$c_k(h) = \sum_{\alpha | k} \alpha \, j_\alpha(h).$$

Here $G = D_6$ and $H = S_2$. The two elements of H are of types s_1^2 and s_2. For the first we have $c_k(h) = s j_s = 2$ for all k; for the second, $s = 2$ and $c_k(h) = 2$ provided k is even. Hence from (6.1) the required number is

$$(6.2) \qquad \tfrac{1}{2}[Z(D_6; 2, 2, 2, \ldots) + Z(D_6; 0, 2, 0, 2, \ldots)].$$

It is easily verified that the required number is 8. Compare this with the number

$$Z(D_6; 2, 2, 2, \ldots) = 13$$

obtained from Pólya's Theorem when the colors are not interchangeable. In general, each necklace will be equivalent to another necklace under interchange of colors, but the fact that there are 13 not
16 necklaces in the Pólya situation shows that there must be exactly
three necklaces equivalent to themselves under interchange of colors.
Moreover this value is exactly that of the term $Z(D_6; 0, 2, 0, 2, ...)$ in
(6.2).

In more recent years several other generalizations of Pólya's
Theorem have appeared. Some of these are discussed elsewhere in
this article, but considerations of space forbid more than a passing
mention of the others. The reader can consult [LloE68] for a Pólya-
type theorem using Dirichlet generating functions, a generalization
of Pólya's Theorem in Theorem 10.1 of [StaR79], and a more general
setting of the theorem in [SolL77]. S. G. Williamson has investigated
the relation between Pólya's Theorem and operator theoretic invariants, and its interpretation in the language of multilinear algebra;
see [WilS70,71,72,73,73a]. This work was continued by White in
[WhiD74,75,75a,75b,75c]. See also the joint paper [WhiD76].

7. Applications to Graphical Enumeration

Pólya's Theorem can be applied in a very straightforward manner to
many problems concerning the enumeration of graphs, and in particular, to the basic problem of counting the number of graphs on a
given number of vertices and edges. We shall assume first that the
graphs have no loops or multiple edges.

If the vertices are labelled with labels 1,2, ..., p then there are $N =$
$\frac{1}{2}p(p - 1)$ pairs of vertices, all distinct, and the problem of finding
the number of such graphs with q edges is simply that of choosing q
elements from a set of N elements. This number is, of course, $\binom{N}{q}$.
If the vertices are unlabelled, however, the problem is less easy and
calls for the use of Pólya's Theorem.

We can regard each pair of vertices as a "box" into which we can
put one of two figures: an "edge", with content 1, or a "non-edge",
with content 0. Thus the figure generating function for this
problem is $1 + x$. The resulting configuration is then a graph, and
its content is the number of edges.

We now need to know the appropriate group G. Since they are not
labelled, we can permute the vertices in any way, that is, by any
element of S_p, the symmetric group of degree p. Each permutation
of the vertices will induce a permutation of the pairs of vertices,
and these permutations form the group that we require. We can
denote it by $S_p^{(2)}$. Pólya's Theorem then gives the configuration
generating function in the form

$$Z(S_p^{(2)}; 1 + x).$$

Now the cycle index of the group $S_p^{(2)}$ can be found for any given p by considering each of the various types of permutations, and finding the type of permutation induced on the edges. In each case the coefficient will be the same as in S_p. For details see [HarF55] and many other references.

Computing the cycle index for general values of p is a little more tricky, though not difficult. It turns out to be the following rather unwieldy expression:

$$(7.1) \qquad \frac{1}{p!} \sum_{(j)} \frac{p!}{\Pi j_k k^{j_k}} \Pi (s_k s_{2k}^{k-1})^{j_{2k}} . \Pi s_{2k+1}^{k j_{2k+1}} . \Pi s_k^{k \binom{j_k}{2}} , \Pi s_{m(r,t)}^{d(r,t) j_r j_t} ,$$

where $m(r,t)$ and $d(r,t)$ are the l.c.m. and g.c.d. of r and t, where the last product is for $1 \leqslant r < t \leqslant p$, and the other products are over the relevant values of k.

Pólya was aware of this application of his theorem but did not publish it. He communicated it to Harary (see [HarF67a] for the details), who set it out along with other results of a similar kind in [HarF55]. These results include the enumeration of directed graphs (by considering the group of permutations of the *ordered* pairs induced by S_p) and multigraphs of strength s, that is, multigraphs in which up to s edges are allowed between two vertices. This requires the use of the figure generating function

$$1 + x + x^2 + \cdots + x^s$$

instead of $1 + x$.

Unwieldy or not, the cycle index given in (7.1) can be used to compute the number of graphs; either by hand if p is small or by computer for the larger values. In 1964 King and Palmer [KinC64] computed the total number of graphs for each value of p up to 24. More recently Stein and Stein [SteM67] enumerated graphs with given numbers of vertices and edges up to $p = 18$.

This work of Harary was not, strictly speaking, the earliest enumeration of unlabelled graphs. The paper by Davis [DavR53] two years earlier is effectively concerned with graphical enumeration, though disguised as a problem in logic, and Slepian [SleD53] had tackled the same problem. Harary, in [HarF67a], references two other early enumerations in unpublished papers by A. M. Gleason and S. Golomb; and J. H. Redfield, who will be discussed later, also anticipated this work much earlier.

A series of four papers by G. W. Ford and others [ForG56,56a,56b, 57] amplified this work by using Pólya's Theorem to enumerate a variety of graphs on both labelled and unlabelled vertices. These included connected graphs, stars (blocks) of given homeomorphic type, and star trees. In addition many asymptotic results were derived. The enumeration of series-parallel graphs followed in 1956 [CarL56], and in that and subsequent years Harary produced variations on the basic graph enumeration problems: line-subgraphs [HarF56], oriented graphs [HarF57], supergraphs of a subgraph [HarF57a], subgraphs between a given graph-subgraph pair [HarF58a], and bicolored graphs [HarF58], which were also enumerated in [ReaR58]. The enumeration of functional digraphs [HarF59] was along the same general lines, and the formula obtained by Harary was simplified by me in 1961 [ReaR61].

A type of problem to which Pólya's Theorem was not applicable was the enumeration of regular graphs. A k-regular graph is one in which every vertex has degree k, i.e., is incident with k edges. This condition means that what is put in one box will influence what is put in others, and under these circumstances Pólya's Theorem in its usual form is not applicable. Using the superposition theorem, developed in 1958 [ReaR58], I was able to make some progress with this problem [ReaR59,60].

In 1960 Harary published a list of 27 unsolved problems in enumerative graph theory [HarF60]. From time to time thereafter he revised this list as some problems were solved and others propounded, but always contriving that the list contained 27 problems [HarF64,67b,70a]. One problem listed in the first of these papers was the enumeration of self-complementary graphs, that is, graphs that are isomorphic to their complement. This is not a Pólya-type problem but yields readily to de Bruijn's generalization, and the enumeration of self-complementary graphs was accomplished in 1963 in [ReaR63]. This paper also enumerated self-complementary digraphs, and disclosed the curious fact that the self-complementary graphs on $4n$ vertices are equinumerous with the self-complementary digraphs on $2n$ vertices. It seems likely that there should be some "natural" one-to-one correspondence between these two sets; but if there is, it has so far eluded discovery. A similar result discovered some years later is that the number of self-complementary graphs on $4n + 1$ vertices is equal to the number of self-complementary relations over $2n$ elements (see [WilD78]).

By way of contrast we can note one problem that persisted throughout the series of Harary's papers on unsolved problems, and which is still unsolved. This is the problem of enumerating identity graphs -- graphs with only the identity automorphism, or, in other words, graphs with no symmetry. This problem has remained intractable as has, a fortiori, the more general problem of enumerating graphs with a given automorphism group. Stockmeyer [StoP71], Sheehan [SheJ68], and others have shed light on this problem, though without coming up with a really practical solution to it.

By this time -- the late 1960s -- it was probably true to say that most of the significant graphical enumeration problems that were amenable to a direct application of Pólya's Theorem or of some generalization of it, had already been solved. New enumerative techniques were needed -- techniques which, while still using Pólya's Theorem, applied it in a more indirect way. We shall look at one such technique, the method of cycle index sums, in a later section.

8. The Work of J. H. Redfield

All this time -- behind the scenes, as it were -- was a paper which, since 1927, had been waiting to be recognized as a remarkable contribution to combinatorial theory. Its title, "The Theory of Group Reduced Distributions", gave no immediate intimation of its content, and the author, J. H. Redfield, was not a well-known mathematician; indeed, this was the only mathematical paper by him published during his lifetime. Nevertheless the paper appeared in a reputable journal (the reference is [RedJ27]), and it is surprising that it went unnoticed for so long; but that was what happened. Not until 1960 did Frank Harary discover and reveal to the combinatorial world the fact that this paper contained within it the gist of much of the enumerative work that had been done since the publication of Pólya's paper. Unlike Pólya's paper, which delved deeply and thoroughly into the applications and ramifications of the new theorems, Redfield's paper often did little more than hint at some of the results he had discovered. Nevertheless, it is clear that he had an equivalent of Pólya's Theorem, and knew how to enumerate unlabelled graphs by numbers of vertices and edges. He also stated and used the superposition theorem. In addition, he shed much light on properties of cycle indexes (which he called group reduction functions) and showed how to work with the sum of the cycle indexes corresponding to a set of combinatorial objects.

The discovery of Redfield's paper had an immediate impact on combinatorial research. Various expository papers appeared, interpreting Redfield's results in modern notation (see, for example, [HarF67c], [FouH63], [FouH66]), and soon the effects of his discovery began to become apparent.

Among other things, Redfield's paper led to a heightened awareness of something that was already beginning to be realized, namely the interrelationship between Pólya's Theorem (and other enumeration theorems) on the one hand, and the theory of symmetric functions, S-functions, and group characters on the other; it helped to show the way to the use of cycle index sums in the solution of hitherto intractable problems; and in a more nebulous way it provided a refreshing new outlook on combinatorial problems.

Redfield had written and submitted a second paper in 1940, but it had been rejected. It remained among his effects after his death in 1944, and by the kindness of his daughter, Mrs. Priscilla Redfield Roe, was retrieved and published in a special issue of the *Journal of Graph Theory* [RedJ84]. This special issue also included a paper by Hall, Palmer, and Robinson [HalI84] interpreting Redfield's work in the light of modern combinatorial mathematics. Further information about Redfield and his work can be found in E. K. Lloyd's biographical article [LloE84] in the same issue of the journal.

The use of cycle index sums, mentioned above, is a powerful technique for enumeration, and a comparatively recent development of Pólya theory. It will be instructive to consider a simple example

of its use, and to that end let us consider the problem of enumerating connected graphs with no vertices of degree 1 -- a problem that
was long considered intractable and not amenable to any form of
Pólya's Theorem.

If we progressively delete vertices of degree 1 from a connected
graph C, until no more such vertices remain, we shall obtain a connected graph F which we can call the "frame" of C. The graph C is
then seen to consist of the graph F at each vertex of which a rooted
tree has been attached, so that the root of the tree is identified with
the vertex of F. Figure 8 shows a typical example.

Suppose we want to enumerate, by number of vertices, connected
graphs whose frame is a given graph F_1. This is a straightforward
Pólya-type problem. The "boxes" are the vertices of F_1, the figures
are rooted trees, and the group is the automorphism group of F_1; call
it A_1. Pólya's Theorem gives $Z(A_1; T(x))$ as the required configuration generating function, where $T(x)$ is the generating function for
rooted trees.

For our present purpose we shall need to retain much more information about these graphs. Specifically, we want to find the sum of
the cycle indexes of their automorphism groups. This is still basically a Pólya-type problem, for which we replace $T(x)$ by the sum of
the cycle indexes of rooted trees. If T denotes the set of rooted
trees, then this cycle index sum can be written $Z(T)$. Note that we
can always recover $T(x)$ from $Z(T)$; for since the sum of the coefficients in the cycle index is 1, we have only to replace each occurrence of s_i by x^i. Each cycle index for a tree on n vertices then reduces to x^n. This result is general and applies to any cycle index
sum.

The required result can be written $Z(A_1; Z(T))$; but since we now
have a process very much akin to that of substituting one cycle
index in another -- the process by which the cycle index of a wreath
product is obtained -- it is usual to use the notation of the wreath
product and write $Z(A_1)[Z(T)]$.

Now suppose we do this with every graph F_1 in the set, F say, of
frames and add the results. We shall obtain a cycle index sum for

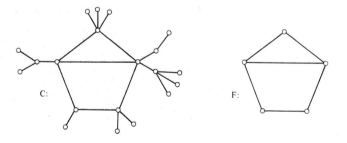

Figure 8. A connected graph, C, and its frame F.

the automorphism groups of all connected graphs; for every such graph that can be obtained by adding rooted trees to a frame. Thus if C denotes the set of connected graphs, we have the equation

$$(8.1) \qquad Z(C) = Z(F)[Z(T)],$$

where $Z(C)$ denotes the sum of the cycle indices of the automorphism groups for connected graphs.

In this equation it is the cycle index sum $Z(F)$ that we do not know. The cycle index sums $Z(T)$ and $Z(C)$ *can* be computed, though the method is not immediately apparent and is a story in itself. Thus equation (8.1) does, in theory, give a means for computing $Z(F)$, but not a very practical one. The equation is the wrong way round, and what is needed is an expression for $Z(F)$ in terms of $Z(C)$ and $Z(T)$. In what was an important breakthrough in this kind of enumeration, R. W. Robinson showed that equation (8.1) could be inverted to give an equation of the form

$$F(x) = Z(C)[M(x)],$$

where $F(x)$ is the required generating function for frames, and where $M(x)$ can be effectively computed.

By means of his "composition theorem", which effects this inversion, Robinson was able to solve other difficult problems, such as that of enumerating unlabelled blocks and acyclic digraphs, in which, as above, the required generating function is defined by an equation in which it appears to be inextricably entangled with other functions. The way in which the composition theorem serves to extricate the required generating function is explained in Robinson's papers [RobR70,70a,73,76]. Hanlon [HanF81] has given a combinatorial interpretation of the function $M(x)$ which appears in the above example.

The light shed by Redfield's paper on the close connection between Pólya theory and symmetric function theory is well illustrated by a particularly simple way of looking at Pólya's Theorem -- one that shows the way to further developments. Suppose the store of figures consists of n distinct figures, as for example with necklace problems using n kinds of beads. The figure generating function is then

$$(8.2) \qquad x_1 + x_2 + x_3 + \cdots + x_n$$

and the configuration generating function is obtained by replacing, in the cycle index, every occurrence of s_r by $x_1^r + x_2^r + \cdots + x_n^r$. But this latter expression, called a power sum, is of common occurrence in the theory of symmetric functions, where it is universally denoted by s_r. It was for this reason that s_r was used in the notation for the cycle index rather than Pólya's f_r.

It follows then that, as was pointed out in [ReaR68], the cycle index and the configuration generating function are one and the same thing, and Pólya's Theorem reduces to the statement of this equivalence. Note that this is not just a special case, due to the particular form of the figure generating function (8.2). If the i-th figure is given a constant n_i, and occurs m_i times, its contribution to the total content is $m_i n_i$, and to enumerate configurations by total content we need only replace x_i in (8.2) by x^{n_i}. Thus (8.2) becomes

$$\sum_{\text{all figures}} x^{n_i},$$

which is the general form for a finite figure generating function in one variable. This result is easily extended to the infinite case, and to the case of several variables. Thus there are good reasons for interpreting the indeterminates s_i in the cycle index as power sums. In this way every cycle index becomes a symmetric function, and the door is then open to the use of the well-developed theory of symmetric functions, especially the theory of group representations. This leads to a fruitful liaison between Pólya theory and many aspects of linear algebra.

It was shown in [ReaR68] that the operation $N(A*B)$ which is required by the superposition theorem is particularly simple if the operands A and B are the symmetric functions known as S-functions (or Schur functions). In fact, for any two S-functions $\{\lambda\}$ and $\{\mu\}$

$$N(\{\lambda\}*\{\mu\}) = 1 \quad \text{if} \quad \lambda = \mu$$

$$\text{and} = 0 \quad \text{if} \quad \lambda \neq \mu,$$

where λ and μ are two partitions of an integer n. Hence if two symmetric functions A and B are expressed as sums of S-functions, say

$$A = \sum_{\lambda} a_\lambda \{\lambda\} \quad \text{and} \quad B = \sum_{\mu} b_\mu \{\mu\},$$

then

$$N(A*B) = \sum_{\rho} a_\rho b_\rho,$$

where the last summation is over all partitions ρ of n. If two cycle indexes A and B can be easily expressed in terms of S-functions (and they often can), then the computation of $N(A*B)$ becomes much simpler. In particular the wreath product $S_n[S_2]$ and certain other related cycle indexes have simple expansions in terms of S-functions.

We have already seen how the cycle index of the wreath product $G[H]$ is obtained by substituting the cycle index $Z(H; s_1, s_2, \ldots)$ of H in the cycle index $Z(G; s_1, s_2, \ldots)$ of G. Now this process of substitution can be extended, in an obvious way, to "substituting" one symmetric function, expressed as a polynomial in the power sums s_i, in another such function. It turns out that the result of this operation is well known in symmetric function theory as a "plethysm", and that what in Pólya's notation would be denoted $A[B]$, is there written as $B \otimes A$ ("B plethys A") (see [LitD50]). In view of the essentially substitutional character of this operation, Pólya's notation can be deemed superior. Much work has been done by algebraists on plethysms, and their labors can be turned to further good use by combinatorialists.

It would take us too far out of our way to explore the territory opened up by this way of looking at Pólya's Theorem, but we can take a quick look at the view and pick out some of the salient features. The cycle index, in its new clothes, is expressible in terms of S-functions; conversely, an S-function can be regarded as a sort of cycle index. The S-function for a partition can be defined by

$$(8.3) \qquad \{\lambda\} = \frac{1}{n!} \sum_{\rho} A_\rho \chi_\rho^\lambda s_\rho,$$

where ρ is a general partition -- corresponding to our previous (j), where s_ρ stands for $s_1^{j_1} s_2^{j_2} ...$, where A_ρ denotes the number of group elements of type ρ, and χ_ρ^λ is a group characteristic. This is called the character weighted cycle index by White [WhiD80]. If λ is the partition (n) then (8.3) is the cycle index of S_n; if λ is (1^n) then (8.3) is $Z(A_n) - Z(S_n)$, that is, the polynomial into which the figure generating function is to be substituted in order to count configurations with no repetitions of figures. White, in the same paper, showed that for general λ, the character weighted cycle index can be used to enumerate sums of weights of column-strict tableaux.

9. Applications to Chemical Enumeration

The enumeration of chemical compounds was of particular interest to Pólya, to the extent that he specified it explicitly in the title of his paper. We have already looked at his achievements with acyclic compounds, and subsequent developments in their enumeration. We now turn to the problem of enumerating cyclic chemical compounds, a problem that is much more difficult and which, in its most general form, does not seem to admit any useful theoretical solutions.

The general problem of chemical enumeration is as follows. Given the number of atoms of each kind that occur in a molecule, determine the corresponding number of possible molecules, either as structural isomers or stereoisomers. Computer programs have been written which construct all such compounds for a given set of atoms, thus determining the number of compounds for any given instance of the problem (see [MasL74] and [TriN83]); but no theoretical solution, say in the form of a generating function, is known. For more on chemical enumeration in general, see [RouD71,72,76].

The corresponding graph theoretical problem is that of determining the number of unlabelled graphs with given numbers of vertices of given degrees, and is equally difficult. Parthasarathy [ParK68] has given an expression for the solution to this problem, and so has Hanlon [HanP79], working from an entirely different angle; but these solutions do not appear to be of practical value in obtaining exact numerical results. A computerized approach to this problem was outlined in [ReaR80] and sufficed to generate the numbers of graphs on 10 vertices according to their degree sequences; but even this method would be impracticable for large graphs.

For this reason such results as have been obtained for chemical enumeration have been for restricted types of compounds. The usual method is to consider compounds which have a common basic structure. Imagine the structural formula of a cyclic compound, and perform as many times as possible the operation of removing an atom of valency (degree) 1, together with its bond, just as we did in forming the frame of a graph in the last section. We shall be left with a structure having no atoms of valency 1. This is illustrated in Figure 9.

The basic structure thus defined can be called the frame of the molecule. Reversing this procedure we can define a class of compounds by taking a given frame, and attaching to each atom which, in the frame, has less than its proper valency, enough acyclic radicals to bring its valency up to the correct value. (Figure 9, taken in reverse, illustrates this process.) Since the generating function for these radicals needs to be known, the range of possible radicals will need to be restricted. A common restriction is to stipulate that they are single atoms; another is that they should be alkyl radicals. Polya studied several problems of this kind, including those for which the frame was the 3-ringed carbon structure of cyclopropane, the benzene ring, and, in [PolG36], the structures of naphthalene, anthracene, pyrene, phenanthrene, and thiophene noted earlier.

In the general problem of this type the figure generating function would be that for the allowable radicals, and the group will be the group of automorphisms of the frame, restricted to those atoms which do not enjoy their full valency within the frame.

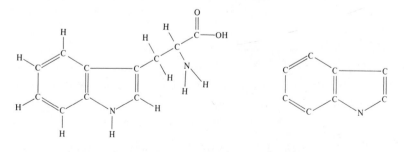

Figure 9. A structural formula and its frame.

It must be admitted that many chemical enumerations that have
appeared in the literature are more in the nature of academic exer-
cises than results of practical interest to the chemist. Thus, for
example, no chemist *really* needs to know that the number of alkanes
having 60 carbon atoms is 22,158,734,535,770,411,074,184 (see
[PerD32]). However, the enumeration of compounds with a given
frame can produce results of practical importance.

An interesting case in point is that of the number of isomers of
porphyrin. The porphyrins are compounds derived from porphin,
which has the structure shown (somewhat simplified) in Figure 10.
In this figure the numerals 1 to 8 indicate the sites at which substitutions may take place. Two kinds of porphyrins are of particular
interest: (a) those in which one kind of radical is placed at four of
these sites and a second, different, radical at each of the other four;
and (b) those in which one kind of radical is placed at four sites,
another kind at two other sites, and yet a third radical at the remaining two. How many possibilities are there for the two cases (a)
and (b)? The famous organic chemist Fischer [FisH66] had stated
that the numbers were 4 and 15, respectively. These figures were
questioned by Blackman [BlaD73] who obtained the values 8 and 33.
These values are incorrect since, as was pointed out by Pilgrim
[PilR74], they should be 13 and 16. Neither of these authors made
use of Pólya's Theorem, but it is an elementary exercise in the use
of the theorem, with $\frac{1}{8}(s_1^8 + 2s_4^2 + 5s_2^4)$ as cycle index, to verify that
the latter are indeed the correct values. In fairness to Fischer it
should be mentioned that his figures were based on a restrictive
assumption concerning the mode of substitution and, given that
assumption, are correct. For the derivation of these results using
Pólya's Theorem, as also those with Fischer's restrictions, see
[AlsB77] and [TapR78]. For similar problems relative to other basic
chemical structures, see for example, [BalA67,70,76a], [FluR84],
[HaiC77], [KenB64], [NouJ79], [RouD75a].

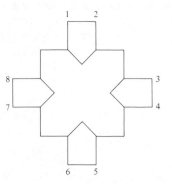

Figure 10. Diagram of the porphin structure.

A particularly comprehensive account of work on this kind of problem is contained in the Ph.D. thesis of R. A. Davidson [DavR77]. Enumerations are carried out for a wide variety of frames, and with various conditions on the nature of the radicals. This thesis also contains a full account of tree enumeration in chemistry. Chemists have been very ready to use Pólya's theory and its generalizations in their field, and much work of this nature has been done -- too much to be described in detail here. The following additional references form a representative selection: [BurJ81,82], [KinR81], [KnoO75], [KorM75], [McLT82].

Again it is worth noting how real-life considerations may complicate an otherwise simple combinatorial problem. In Section 1, when we looked at derivatives of benzene, we tacitly assumed that the benzene ring was flat -- a plane hexagon. It so happens that for benzene this assumption is justified; but the same is not true for cyclohexane, which also has a ring of six carbon atoms. This ring is "buckled", so to speak, and this buckling must be taken into account. For the effect that this has on the enumeration of its derivatives, see [LeoJ75,77], [FluR76], and [SlaZ81, section b].

The use of Pólya's Theorem in a specialized context such as the above, has led to the extension of the theorem along certain useful lines. One such derivation pertains to the situation where the boxes are not all filled from the same store of figures. More specifically, the boxes are partitioned into a number of subsets, and there is a store of figures peculiar to each subset. To make sense of this we must assume that no two boxes in different subsets are in the same orbit of the group in question. A simple extension of Pólya's Theorem enables us to tackle problems of this type. Instead of the cycle index being a function of a single family of variables, the s_i, we have other families of variables, one for each subset. An example from chemical enumeration will make this clear.

Consider derivatives of cyclopropane (Figure 11) obtained as follows. Each hydrogen atom may be left unchanged or can be

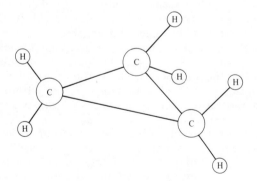

Figure 11. Cyclopropane.

replaced by a fluorine atom (F) or an iodine atom (I); each carbon atom can be left unchanged or replaced by an atom of radioactive carbon. We seek the number of distinct derivatives. Here we have a set of six boxes (the locations of the hydrogen atoms) for which the figures are H, F, or I; and we have a set of three boxes (the carbons) for which the figures are "normal carbon" or "radioactive carbon". To keep these two sets of boxes distinct we shall retain the usual s_i for the first set, and use a new family of variables, t_i, for the second set.

Consider the group element which rotates the molecule about a median of the central triangle. It interchanges the hydrogen locations in pairs, leaves one carbon location fixed, and interchanges the other two. Hence for this group element we write down the monomial $s_2^3 t_1 t_2$. In this way we write down, for the whole group, what Davidson calls the compound cycle index, namely

$$\tfrac{1}{6}(s_1^6 t_1^3 + 2s_3^2 t_3 + 3s_2^3 t_1 t_2).$$

The figure generating function for the two sets are $H + F + I$ and $x + y$ (in an obvious notation), and a complete description of all distinct possibilities is given by the configuration generating function

$$\tfrac{1}{6}\{(H+F+I)^6(x+y)^3 + 2(H^3+F^3+I^3)^2(x^3+y^3)$$

$$+ 3(H^2+F^2+I^2)^3(x+y)(x^2+y^2)\}.$$

If we want only the total number of possibilities, we put $H = F = I = x = y = 1$ and obtain

$$\tfrac{1}{6}\{3^6 2^3 + 2.3^2.2 + 3.27.2.2\} = 1032.$$

This trick of manipulating two or more families of variables simultaneously can be used also with applications of the superposition theorem. This idea seems to have been discovered independently by several workers; for an example see [Mull69], where it is used to enumerate a class of regular digraphs.

Another development that has stemmed from chemical enumeration arises from the need to find the number of substituted compounds for which the numbers of substituents of the various kinds are given beforehand. To obtain such a result without at the same time obtaining the results for *all* numbers of substituents, Ruch and others [RucE70,72,83] developed a method of "double cosets" which they are able to use to good effect. It turns out, however, that this result is in effect the superposition theorem used, as described earlier, to find particular terms in the configuration generating function without computing the whole function. It therefore goes back essentially to the work of Redfield. Chemical enumerators have been quick to realize the significance of Redfield's work and its applicability to chemical problems. For example, we have seen how Pólya's Theorem enables us to enumerate isomers. Let us go one step further, and consider the process by which one isomer can change into another by rearranging the radicals that are attached to a given frame. The problem of counting the number of ways this can happen -- the number of *permutational isomerizations* -- has been addressed by Klemperer [KleW72], Klein and Cowley [KleD75], Nourse [NouJ77], Davidson [DavR81], and others. The method is clearly described in [LloE85], where it is shown that this reduces to an application of the superposition theorem.

A very recent paper in the field of chemical enumeration is [GarJ85], which describes a computer program that carries out the tedious work of substituting the appropriate figure generating functions into the cycle index of the group of automorphisms of a frame, collecting up the resulting terms, and grouping them according to their partitions. In this way extensive calculations are performed for the permutation isomers of nonrigid pentane, adamantane, dodecahedane, and rigid and nonrigid biphenyl. A copy of this paper was sent to Pólya shortly before his death in September 1985.*

10. Back to Burnside

Powerful though Pólya's Theorem undoubtedly is, it is not difficult to formulate problems which, though superficially very similar to the general Pólya-type problem, cannot be solved by use of the theorem. We have already seen how some problems of this kind led to generalizations of Pólya's Theorem, such as de Bruijn's Theorem, the power group enumeration theorem, and the superposition theorem, but even these theorems have their limitations. For some problems not amenable to solution by any of these results, it is sufficient to fall back on the result from which they all stem, namely Burnside's Lemma.

For example, one of the important requirements of a Pólya-type problem is that the placing of the figures in the boxes should be independent, that is, that the placing of the figure in one box should not be restricted or influenced by what has been placed in another box. If this is not so -- if there is interaction between the boxes, so to speak, then the problem takes on an entirely different aspect.

Consider the following necklace problem. How many necklaces can be made from six beads, with three colors of beads available, subject to the restriction that no two adjacent beads are to be the same color? Without the restriction the problem is a straight Pólya-type problem of a kind that we have already met, and the solution is $Z(D_6; 3,3, ...) = 92.$ With the restriction, however, the problem no

---

*I am indebted to G. L. Alexanderson of Santa Clara University for this piece of information.

longer fits Pólya's Theorem, but can be solved by using Burnside's Lemma. For each permutation of the group D_6 we find how many necklaces are invariant under the action of that permutation. For the identity we have the problem of coloring the vertices of a fixed hexagon, H, in three colors so that no two adjacent vertices are colored alike. This is a straightforward graph coloring problem, and the solution is the value, at $x = 3$, of the chromatic polynomial $P_H(x)$ of H. Since $P_H(x) = (x-1)^6 - (x-1)$, we have $2^6 + 2 = 66$ as the number of colorings. (For details on this see [ReaR68b].)

Now the two permutations of cycle-type s_6 leave no colorings invariant since they map vertices onto adjacent vertices. The two rotations of type s_3^3 require alternate vertices to be similarly colored and different from their neighbors. This is equivalent to coloring the graph consisting of two vertices and one edge, and the number of colorings is clearly 6. In general, for a given permutation, the colorings are those of the graph obtained from H by identifying vertices in the same orbit, with suppression of double edges if these should occur. For the permutations of type s_3^2 the graph is a triangle with six colorings; while for each of the three reflections of type $s_1^2 s_2^2$ the graph is the path of length 3, with 24 colors. The other reflections, of type s_2^3, map some vertices on to their neighbors, and thus do not give rise to any proper colorings.

Thus, applying Burnside's Lemma, we find the number of necklaces to be

$$\tfrac{1}{12}[66 + 2 \times 6 + 6 + 3 \times 24] = 13.$$

This method can be used more generally to find what can be called the "unlabelled chromatic polynomial" of a graph -- giving the number of ways of coloring in x colors the vertices of a graph when two colorings are equivalent if one is converted to the other by an automorphism of the graph.

Another type of problem to which Pólya's Theorem is not directly applicable occurs when the action of the group G which permutes the boxes has an effect on the contents of the boxes. This effect can be seen in the problem of enumerating regular dissections of a polygon, or equivalently of tree-like "clusters" obtained by joining together regular polygons. This problem is studied in [HarF70]. At one stage in the solution of the problem it is necessary to enumerate configurations consisting of a "root" hexagon, to alternate sides of which are attached clusters rooted at an edge. Figure 12 gives the general idea of what is happening.

Now this is at first sight a straight Pólya-type problem, with three boxes, permutable by the group S_3 (since reflections are allowed) and with the clusters as figures. Unfortunately those elements of S_3 that reflect the root hexagon also replace each figure by its mirror image. Thus the action of the group can change the figures in the boxes. This might not be so bad if the figures were *always* changed, but if the figure happens to be symmetrical relative to that group

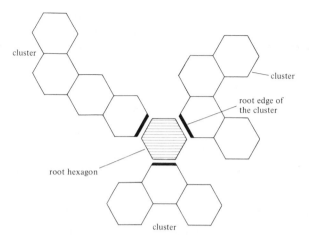

Figure 12. Formation of a rooted polyhex.

operation then it is *not* changed. This is a complication which neither Pólya's nor de Bruijn's theorems can handle. For although the latter theorem allows the possibility of permutations of the figures, these permutations are completely separate from the permutations of the boxes, whereas what we have here is a group that acts on the boxes and the figures in an interrelated way.

We can solve this problem by going back to Burnside's Lemma and counting, for each group element, the number of configurations that are invariant under it. This will require more information about the figures in that we now need to know which figures have symmetry and which do not. This approach was used in enumerating tree-like polyhexes [HarF70] and several variations of the problem of dissecting a polygon into smaller polygons by chords (see [HarF75a], [ReaR78]). The enumeration of stack polytopes [HerF82] presented the same problem, and yielded to the same method of solution.

The type of problem just mentioned is of great importance in the enumeration of chemical isomers when the situation of a molecule in three-dimensional space is heeded, and leads to the consideration of "chirality".

An object is said to be chiral if it cannot be brought into coincidence with its mirror image by means of translation and rotation. A right-hand glove, for example, is chiral since its reflection is a left-hand glove, which is a distinct object under translation and rotation.

Consider a methane molecule CH_4, and suppose that some or all of its hydrogen atoms are replaced by some other monovalent atom. If the atoms attached to the carbon are all different, that is, the carbon atom is asymmetric, the resulting molecule is chiral and exists in two so-called enantiomorphic forms -- mirror images of each other. (For further information on chirality see the interesting expository paper [PreV76]).

If we apply Pólya's Theorem to this problem with figure generating function $w + x + y + z$ (for four possible substituents) and with S_4 as the group, thus allowing all possible permutations of the substituents, including reflections, then we obtain the configuration generating function $Z(S_4; w + x + y + z)$. To find the total number of configurations we put $w = x = y = 1$ and obtain $Z(S_4; 4,4,4,...) = 35$.

If we disallow reflections, thus heeding the fact that the molecule exists in three-dimensional space, we must use the alternating group A_4 instead of S_4. In this case, as the reader can verify, the total number of configurations becomes 36. This shows that in the latter case there are two distinct configurations which become equivalent if reflection is allowed; these are precisely the two configurations in which the substituents are all different.

When there are many carbon atoms which might be asymmetrical, the solution is more complicated. Nevertheless, by methods that basically rely on Pólya's Theorem, enumeration of compounds taking chirality into account can be carried out. For alkanes and monosubstituted alkanes see the paper [BalA76]; for the chiral alkanes with some restrictions see [QuiL77,79]. See also [HarF75], [PalE77], and [WorN81] for other problems in which chirality appears.

The problem that we noted above with clusters appears also in chemical enumeration when we consider compounds formed by attaching radicals which may be chiral or achiral to a frame which is achiral. In this case, too, Pólya's Theorem cannot be used, but the problem can be solved by the appropriate use of Burnside's Lemma. It is also amenable to the methods of Redfield, as shown in [DavR81] and [LloE85].

However, just as Pólya's Theorem is useful even though it is essentially a particular application of Burnside's Lemma to a standard, though very general, type of combinatorial problem, so it is useful for chemical enumerators to have a theorem that embodies in some simple form the method of solution for the type of problem just mentioned. An extension of Pólya's Theorem along these lines has been published by J. E. Leonard [LeoJ77]. More recently, Hässelbarth [HasW84] has given a theorem which not only covers the case where there is a group action that acts both on the domain and the range of the functions being considered, but also introduces another generalization, namely a relaxation of the condition, required in previous theorems, that the weight should be constant over the orbits.

11. Asymptotics

The three specific topics explicitly mentioned in the title of Pólya's paper, namely Groups, Graphs, and Chemical Compounds, each has a section to itself following the Introduction. A fourth section, equally important, deals with the determination of asymptotic results for the kinds of combinatorial problems discussed in the earlier sections. Pólya brought his great analytical skill to bear on these asymptotic problems, and insofar as this work has provided much inspiration and incentive for later mathematicians, a brief account of some of the research springing from it is in order.

Pólya considered eight different kinds of trees, listed below, together with the symbols for the number of such trees with n vertices.

(i)   Free trees τ_n

(ii)   Rooted trees T_n

(iii)   Free trees with degrees 1 or 4 (alkanes) ρ_n

(iv)   Rooted trees -- degrees 1 or 4 (alkyl radicals) R_n

(v)   As (iii) but counting stereoisomers σ_n

(vi)   As (iv) but counting stereoisomers S_n

(vii)   Alkanes with no asymmetric carbon atoms κ_n

(viii) Alkyl radicals with no asymmetric carbon atoms Q_n.

He established inequalities between these numbers, as follows:

$$1 \leqslant \kappa_n \leqslant \rho_n \leqslant \sigma_n; \qquad \rho_n \leqslant \tau_n ;$$

$$1 \leqslant Q_n \leqslant R_n \leqslant S_n \leqslant \frac{1}{n} \binom{3n}{n-1}; \qquad R_n \leqslant T_n \leqslant \frac{1}{n} \binom{2n-2}{n-1};$$

$$\rho_n \leqslant R_n \leqslant n\rho_n; \qquad \sigma_n \leqslant S_n \leqslant n\sigma_n; \qquad \sigma_n \leqslant T_n \leqslant n\tau_n ,$$

and between the corresponding radii of convergence of the generating functions for τ_n, ρ_n, σ_n, and κ_n, namely

$$\frac{4}{27} < \sigma < \rho < \kappa < 1; \qquad \frac{1}{4} < \tau < \frac{1}{e}$$

in an obvious notation.

From these Pólya used lemma (4.36) of his paper to deduce the following asymptotic results:

$$\tau_n \approx \tau^{-n} n^{-5/2}; \qquad T_n \approx \tau^{-n} n^{-3/2}$$

$$\rho_n \approx \rho^{-n} n^{-5/2}; \qquad R_n \approx \rho^{-n} n^{-3/2}$$

$$\sigma_n \approx \sigma^{-n} n^{-5/2}; \qquad S_n \approx \sigma^{-n} n^{-3/2}$$

$$\kappa_n \approx \kappa^{-n}; \qquad Q_n \approx \kappa^{-n},$$

where $\asymp$ denotes asymptotic proportionality, that is, $a_n \asymp b_n$ means that $\lim_{n \to \infty} a_n/b_n$ = an unspecified constant.

Polya also obtained asymptotic estimates for a number of other problems, such as multiply-substituted alkanes, ethane derivatives, and cycloparaffins.

Further progress in asymptotic tree enumeration was made by Otter, who in [OttR48] considered the problem of rooted and unrooted trees with maximum degree m. Having enumerated unrooted trees by the method already described, he proceeded to derive asymptotic estimates, and after ten pages of analysis arrived at a number of results, of which the following is typical:

$$(11.1) \qquad \tau_n = \frac{\beta^3 \tau^{9/2}}{4\sqrt{\pi}} \tau^{-n} n^{-5/2}$$

where $\beta = 7.924780$ and $\tau = 0.3383219$. This supplies the constant in Polya's result. Its value, on the right-hand side of (11.1), is 0.5349485.

The work of Polya and Otter set the pattern for asymptotic tree enumeration in that their methods were applicable to many problems. Ford and Uhlenbeck [ForG56b], for example, obtained asymptotic numbers for star trees, and further examples of asymptotic enumeration appear in [HarF73] and [BenE74a] (see also [GorM75]). A general procedure for tree asymptotics emerged that was crystallized, so to speak, by Harary, Robinson, and Schwenk in [HarF75b] which presents a 20-step algorithm for the asymptotic enumeration of a large variety of trees.

Polya did not deal with asymptotic enumeration of graphs in his paper, but he was aware of the basic results, namely:

$$(11.2) \qquad g_{p,q} \sim \binom{\frac{1}{2}p(p-1)}{q} / p!$$

and

$$(11.3) \qquad g_p \sim 2^{p(p-1)/2} / p! \, ,$$

where $g_{p,q}$ is the number of unlabelled graphs on p vertices and q edges, and g_p is the total number on p vertices. Since $2^{p(p-1)/2}$ is the number of labelled graphs and $p!$ is the number of ways of labelling a graph having no symmetry, (11.3) embodies the general idea that "most graphs have no symmetry", that is, that almost all graphs have only the identity automorphism. The same interpretation holds for (11.2) provided q is not near its extreme values. These results can be found in the papers [RidR51,53] to which Polya contributed substantially.

Later Oberschelp [ObeW67] obtained the improved result:

$$g_p \sim \frac{2^{p(p-1)/2}}{p!} \left\{ 1 + p(p-1) \left[\frac{1}{2} \right]^{p-1} + O\left[\frac{p^4}{4^p} \right] \right\}.$$

For some related results see [LupO59] and [LisV69].

The vague statement "provided q is not near its extreme values" needs clarifying. Pólya (unpublished) proved that (11.2) holds for $|2q - N| = O(p)$, where $N = p(p-1)/2$; Oberschelp [ObeW67] was able to obtain the condition $|2q - N| < .84p^{3/2}$, and E. M. Wright made further progress, showing that if

$$\mu = 2 \min[q, N - q]/p - \log p,$$

then the following theorems hold.

(a)   If $\mu \to \infty$, almost all graphs have no symmetry;
(b)   if $\mu \leqslant 0$, almost all graphs have symmetry; and in fact
(c)   if $\mu \leqslant 0$ and R is any integer, then almost all graphs have at least R automorphisms.

All these results are found in [WriE74], but the first appeared earlier in [WriE70].

Previously Wright had studied asymptotic relations between the coefficients in two generating functions related by the formula equation

$$(11.4) \qquad \sum_{n \geqslant 0} G_n x^n / n! = \exp \left[\sum_{n \geqslant 0} C_n x^n / n! \right]$$

which is the equation between the numbers G_n of graphs and the numbers C_n of *connected* graphs on n labelled vertices. He also studied the corresponding result for unlabelled graphs, namely

$$(11.5) \qquad \sum_{n \geqslant 0} g_n x^n = \prod_{r \geqslant 1} (1 - x^r)^{-c_r}.$$

In the case of graphs, a rough interpretation of Wright's results is that "almost all graphs are connected", but the theorems embodied in his papers [WriE67,68] are more general than that. Not only do they not specifically deal with graphs, they also give precise conditions under which the ratio C_n/G_n and c_n/g_n tend to 1, for any sets of coefficients related by (11.4) or (11.5). For a variety of other asymptotic problems studied by Wright, see [WriE69,70a,72,74a].

It is often true that the enumeration of labelled graphs is easy, but not always. Regular labelled graphs of degree r, for example, can be enumerated, but only with difficulty. Gould and Jackson [Goul83] have given a general method for doing so, but their methods yield differential equations in the generating functions, and tend to be rather unwieldy.

Asymptotic results for labelled regular graphs are more tractable. Wormald [WorN78] showed that the number of r-regular graphs on p labelled vertices is asymptotic to

$$\frac{(rp)! \; e^{(1-r^2)/4}}{(r!)^n 2^q q!},$$

a result also obtained by Bollobas [BolB80]. This is subsumed in a more general result of Bender and Canfield [BenE78] for the asymptotic number of labelled graphs with given degree sequence.

To conclude this section, we note quickly a few other asymptotic enumerations. In [PalE70] Palmer found an asymptotic estimate for the number of self-complementary graphs and digraphs. Robinson [RobR76] and Stanley [StaR73] enumerated acyclic digraphs, and obtained the asymptotic estimate

$$A \; p! \; 2^{p(p-1)/2} \; \sigma^p \; ,$$

where $A = 1.741061$ and $\sigma = 0.6720075$ (see also [LisV75]). A subsequent paper [BenE86] extends this to asymptotic values for the number $A_{p,q}$ of acyclic digraphs on p vertices and q arcs.

12. Miscellaneous Applications of Polya's Theorem

The two specific areas of research in which Pólya's Theorem has been most extensively applied are graphical and chemical enumeration, a fact which Pólya clearly foresaw in his choice of title. Applications in other fields are far from rare, however, and it is fitting to give a brief account of a few such uses of the theorem.

In the realm of logic, Pólya's Theorem has been used to enumerate the number of Boolean functions under various conditions. An early paper by Pólya himself [PolG40] counted the number of essentially different propositions of n statements, and showed that the problem is equivalent to that of coloring the vertices of a hypercube. Many more problems of this type were studied by Harrison in the early 1960s. In a typical problem of this type, Harrison counts the number of classes of Boolean functions under complementation of variables, permutations of variables, and complementation and permutation together (see [HarM63a]). For other such problems see [HarM63,64], the latter of which makes use of de Bruijn's generalization of Pólya's Theorem. For an extensive overview of this whole topic see [HarM71]. Harrison also used Pólya's Theorem to count finite automata [HarM65] and certain binary matrices [HarM73]. For automata, see also [HarF67d].

Applications of Pólya's Theorem in physics are quite numerous, though for the most part they are effectively graph enumeration problems. Thus in one of the earliest applications of the theorem, Riddell [RidR51] studied various enumeration problems in the context of a problem in physics and obtained, among other results, the numbers of unlabelled graphs with given numbers of vertices and edges. His subsequent paper with Uhlenbeck [RidR53] was on the same lines and dealt with graphs arising in statistical mechanics. Statistical mechanics has provided other opportunities for the use of Pólya techniques (see for example [LarA68]).

Graphs of a special type occur in the study of kinetic structures. Their vertices are links and their edges, of two kinds, represent turning pairs and gear pairs of links. These graphs have no isolated vertices, no circuits of turning edges, and the property that the removal of the geared edges leaves the graph connected. In [BucR70] Pólya's Theorem was used to derive a set of graphs from which, by a hand calculation, these graphs could be extracted. In this connection see also the discussion in [FreF67] where Freudenstein shows how Pólya's Theorem can be effectively used to carry out a graphical enumeration that L. S. Woo had performed by *ad hoc* methods. Mechanisms of a different kind are dealt with from an enumerative point of view in [KopF75], giving rise to a problem on 2-coloring the vertices of a graph. See also [MurF55] for an application to yet another kind of mechanism.

We have already seen many applications of Pólya theory in the enumeration of isomers of organic chemical compounds, but there are other applications of the theory in chemistry. It is occasionally used in the investigation of crystal structure as in [MooP76,76a], [BroC79], [McLT81], and [BerG83]. It is of use in the study of nuclear magnetic resonance and NMR spectroscopy, as in [JesJ73], for instance, and in several papers by Balasubramanian [BalK80a,82, 82b,83,83a]. See also this same author's papers [BalK79a,81,84] for other applications in theoretical chemistry.

To round off this section we note a few unusual applications of Pólya's Theorem; an application to telecommunications network [CatK75], and one to the enumeration of Latin squares [JucA76]. In pure mathematics there is an application in number theory [ChaC82], and one to the study of quadratic forms [CraT80], being the enumeration of isomorphism types of Witt rings of fields. Finally, we note a perhaps unexpected, but quite natural, application in music theory to the enumeration of chords and tone rows for an n-note scale [ReiD85]. In the latter paper it is shown that for the usual chromatic scale of 12 semitones there are 80 essentially different 6-note chords, and 9,985,920 different tone rows.

REFERENCES

AlsB77   Alspach, B., Aronoff, S. Enumeration of structural iso-
          mers in alicyclic hydrocarbons and porphyrins. Canad.
          J. Chem. 55 (1977) 2773-2777.

BalA67   Balaban, A. T., Harary, F. Rev. Roumaine Chim. 12
          (1967) 1511.

BalA70   Balaban, A. T., Farcasiu, D., Harary, F. Chemical
          graphs IX Isotope-isomerism of multiply-labelled
          compounds. J. Labelled Comp. 6 (1970) 211-223.

BalA76   Balaban, A. T., Harary, F., Robinson, R. W. The number
          of chiral and achiral alkanes and monosubstituted
          alkanes. Tetrahedron 32 (1976) 335-367.

BalA76a   Balaban, A. T. Enumeration of Cyclic Graphs. Chap. 5.
          Chemical applications of graph theory (A. T. Balaban, ed).
          Academic Press 1976.

BalK79   Balasubramanian, K. Enumeration of interval rotation
          reactions and their reaction graphs. Theoret. Chim. Acta.
          53 (1979) 129-146.

BalK79a   Balasubramanian, K. A generalized wreath product
          method for the enumeration of stereo and position
          isomers of polysubstituted organic compounds. Theoret.
          Chim. Acta 51 (1979) 37-54.

BalK80   Balasubramanian, K. Graph theoretical characterization
          of NMR groups, non-rigid nuclear spin species and the
          construction of symmetry adapted NMR spin functions. J.
          Chem. Phys. 73 (1980) 3321-3337.

BalK81   Balasubramanian, K. Symmetry simplifications of spare
          types in configuration int by orbital degeneracy.
          Internat. J. Quantum Chem. 20 (1981) 1255-1271.

BalK82   Balasubramanian, K. Topological and group theoretical
          analysis of dynamics nuclear magnetic resonance spectro-
          scopy. J. Phys. Chem. 86 (1982) 4668-4674.

BalK82a   Balasubramanian, K. Computer-assisted enumeration of NMR signals. J. Magn. Res. 48 (1982) 165-177.

BalK83   Balasubramanian, K. Operator and algebraic methods for NMR spectroscopy II. NMR projection operations and spin functions. J. Chem. Phys. 78 (1983) 6369-6376.

BalK83a   Balasubramanian, K. Operator and algebraic methods for NMR spectroscopy I. Generation of NMR spin species. J. Chem. Phys. 78 (1983) 6358-6368.

BalK84   Balasubramanian, K. Comments on graph theoretical analysis of NQR spectra of crystals. Theor. Chem. Acta 64 (1984) 415-419.

BenE74   Bender, E. A., Canfield, E. R. The asymptotic number of labelled graphs with given degree sequence. J. Combinatorial Theory Ser. A 24 (1978) 296-307.

BenE74a   Bender, E. A. Asymptotic methods in enumeration. SIAM Revue 16 (1974) 485-515.

BenE86   Bender, E. A., Richmond, R., Robinson, R. W., Wormald, N. C. The asymptotic number of acyclic digraphs. To appear in Combinatorica.

BerC58   Berge, C. Theorie des Graphes et ses Applications. Dunod, Paris 1958.

BerG83   Berces, G., Kovacs, I. Vacancies and vacancy complexes in binary alloys. Phil. Mag. A. 48 (1983) 838-901.

BlaD73   Blackman, D. Isomers of the porphyrins. J. Chem. Education 50 (1973) 258-259.

BlaC31   Blair, C. M., Henze, H. R. The number of structurally isomeric alcohols of the methanol series. J. Amer. Chem. Soc. 53 (1931) 3042-3046.

BlaC31a   Blair, C. M., Henze, H. R. The number of isomeric hydrocarbons of the methane series. J. Amer. Chem. Soc. 53 (1931) 3077-3085.

BlaC32   Blair, C. M., Henze, H. R. The number of stereoisomeric and non-stereoisomeric mono-substitution products of the paraffins. J. Amer. Chem. Soc. 54 (1932) 1098-1106.

BlaC32a   Blair, C. M., Henze, H. R. The number of stereoisomeric and non-stereoisomeric paraffin hydrocarbons. J. Amer. Chem. Soc. 54 (1932) 1538-1545.

BlaC33   Blair, C. M., Henze, H. R. The number of structurally isomeric hydrocarbons of the ethylene series. J. Amer. Chem. Soc. 55 (1933) 680-686.

BlaC34   Blair, C. M., Henze, H. R. The number of structural isomers of the more important types of aliphatic compounds. J. Amer. Chem. Soc. 56 (1934) 157.

BolB80   Bollobás, B. A probabilistic proof of an asymptotic formula for the number of labelled regular graphs. Europ. J. Comb. 1 (1980) 311-316.

BroC79   Brown, C., Heaton, B. T., Towl, A. D. C., Chini, P.,
            Fumagalli, A., Longoni, G. Stereochemical non-rigidity of
            a metal polyhedron; carbon-13 and platinum-195 Fourier
            transform nuclear magnetic resonance spectra of
            $[t_n(CO)_{2n}]^{2-}$ (n = 3,6,9,12 or 15). J. Organomet. Chem. 181
            (1979) 233-254.

BucR70   Buchsbaum, R., Freudenstein, F. Synthesis of kinematic
            structures of geared kinematic chains and other
            mechanisms. J. Mechanisms 5 (1970) 357-392.

BurJ81   Burdett, J. K., McLarnan, T. J. A study of the arsenic,
            black phosphorus and other structures derived from rock
            salt by bond-breaking processes. I. Structural enumera-
            tion. J. Chem. Phys. 75 (1981) 5764-5773.

BurJ82   Burdett, J. K. Predictions of the structure of complex
            solids. Adv. Chem. Phys. 49 (1982) 47-113.

BurW11   Burnside, W. Theory of groups of finite order (second
            edition). Cambridge University Press, Cambridge, 1911.

CarL56   Carlitz, L., Riordan, J. The number of labeled two
            terminal series-parallel networks. Duke Math. J. 23 (1956)
            435-445.

CatK75   Cattermole, K. W. Graph theory and the telecommunica-
            tions network. Bull. Inst. Math. and Applications 11
            (1975) 94-106.

CayA57   Cayley, A. On the theory of the analytical forms called
            trees. Phil. Mag. 13 (1857) 172-176.

CayA59   Cayley, A. On the theory of the analytical forms called
            trees II. Phil. Mag. 18 (1859) 373-378.

CayA74   Cayley, A. On the mathematical theory of isomers. Phil.
            Mag. 67 (1874) 444.

CayA75   Cayley, A. On the analytical forms called trees with ap-
            plications to the theory of chemical compounds. Rep. Brit.
            Ass. Adv. Sci. (1875) 257-305.

CayA77   Cayley, A. On the number of univalent radicals C_nH_{2n+1}.
            Phil. Mag. Series 5, 3 (1877) 34-35.

CayA81   Cayley, A. On the analytical forms called trees. Amer.
            Math. J. 4 (1881) 266-269.

CayA89   Cayley, A. A theorem on trees. Quar. J. Pure Appl. Math.
            23 (1889) 376-378.

ChaC82   Chao, C-Y. Generalizations of theorems of Wilson,
            Fermat and Euler. J. Number Theory 15 (1982) 95-114.

CofD33   Coffman, D. D., Henze, H. R., Blair, C. M. The number of
            structurally isomeric hydrocarbons of the acetylene series.
            J. Amer. Chem. Soc. 55 (1933) 252-253.

ComL70   Comtet, L. Analyse Combinatoire v. 2. Presses Univer-
            sitaires de France 1970.

CraT80   Craven, T. C. An application of Pólya's theory of count-
            ing to an enumeration problem arising in quadratic form
            theory. J. Combinatorial Theory A 29 (1980) 174-181.

DavR53   Davis, R. L. The number of structures of finite relations.
          Proc. A.M.S. 4 (1953) 486-495.
DavR77   Davidson, R. A. Unified Combinatorial Molecular Stereo-
          analysis. Ph.D. Thesis. The Pennsylvania State Univer-
          sity, May 1977.
DavR81   Davidson, R. A. Isomers and isomerization: Elements of
          Redfield's combinatorial theory. J. Am. Chem. Soc. 103
          (1981) 312-314.
deBN59   de Bruijn, N. G. Generalization of Pólya's fundamental
          theorem in enumerative combinatorial analysis. Koninkl.
          Nederl. Akad. Wetensch. A62 = Indag. Math. 21 (1969)
          59-69.
deBN63   de Bruijn, N. G. Enumerative combinatorial problems
          concerning structures. Nieuw Archief voor Wisk. (3) XI
          (1963) 142-161.
deBN64   de Bruijn, N. G. Pólya's theory of counting, Chap. 5 of
          Applied Combinatorial Mathematics (E. F. Beckenbach, ed.)
          Wiley, New York, 1964.
deBN67   de Bruijn, N. G. Color patterns that are invariant under
          a given permutation of the colors. J. Comb. Theory 2
          (1967) 418-421.
deBN69   de Bruijn, N. G. Enumeration of tree-shaped molecules.
          Recent Progress in Combinatorics (W. T. Tutte, ed.)
          Academic Press (1969) 59-68.
deBN71   de Bruijn, N. G. Recent developments in enumeration
          theory. Actes du Congres Intern. des Math. (Nice 1970)
          Gauthier-Villars, Paris 1971. tome 3, 193-199.
deBN71a   de Bruijn, N. G. A survey of generalizations of Pólya's
          enumeration theorem. Nieuw Archief voor Wisk. (2) XIX
          (1971) 89-112.
deBN72   de Bruijn, N. G. Enumeration of mapping patterns. J.
          Comb. Theory Ser A 12 (1972) 14-20.
FisH66   Fischer, H. Nobel Lectures. Chemistry 1922-41. Nobel
          Foundation, Elsevier, Amsterdam 1966.
FluR76   Flurry, R. L. On the apparent symmetry of cyclohexane.
          J. Phys. Chem. 80 (1976) 777-778.
FluR84   Flurry, R. L. Isomer counting for fluctional molecules. J.
          Chem. Educ. 61 (1984) 663-665.
ForG56   Ford, G. W., Uhlenbeck, G. E. Combinatorial problems in
          the theory of graphs I. Proc. Nat. Acad. Sci. 42 (1956)
          122-128.
ForG56a   Ford, G. W., Norman, R. Z., Uhlenbeck. G. E. Combina-
          torial problems in the theory of graphs II. Proc. Nat.
          Acad. Sci. 42 (1956) 203-208.
ForG56b   Ford, G. W., Uhlenbeck, G. E. Combinatorial problems in
          the theory of graphs III. Proc. Nat. Acad. Sci. 42 (1956)
          529-535.

ForG57   Ford, G. W., Uhlenbeck, G. E. Combinatorial problems in the theory of graphs IV. Proc. Nat. Acad. Sci. 43 (1957) 163-167.

FouH63   Foulkes, H. O. On Redfield's group reduction functions. Canad. J. Math. 15 (1963) 272-284.

FouH66   Foulkes, H. O. On Redfield's range correspondences. Canad. J. Math. 18 (1966) 1060-1071.

FreF67   Freudenstein, F., Woo, L. S. Type synthesis of plane linkages. J. Eng. Ind. 89 (1967) 159-172.

GarJ85   Garavelli, J. S., Leonard, J. E. Improvements in the computer enumeration of permutation isomers. Computers and Chemistry 9 (1985) 133-147.

GarA71   Garsia, A. A presentation of the enumeration theory of Polya and de Bruijn. Analysis Seminar Notes. Univ. of California, San Diego, 1971.

GorM75   Gordon, M., Kennedy, J. W. The counting and coding of trees of fixed diameter. SIAM J. Appl. Math. 28 (1975) 376-398.

GouI83   Goulden, I. P., Jackson, D. M., Reilly, J. W. The Hammond series of a symmetric function and its application to P-recursiveness. SIAM J. Alg. Disc. Math. 4 (1983) 179-193.

HaiC77   Haigh, C. W. The combinatorial enumeration of mesoionic heterocyclic annulenes. MATCH 3 (1977) 87-96.

HalI84   Hall, I. J., Palmer, E. M., Robinson, R. W. Redfield's last paper in a modern context. J. Graph Theory 8 (1984) 225-240.

HanP79   Hanlon, P. Enumeration of graphs by degree sequence. J. Graph Theory 3 (1979) 295-299.

HanP81   Hanlon, P. A cycle-index sum inversion theorem. J. Comb. Theory A30 (1981) 248-269.

HarF55   Harary, F. The number of linear, directed, rooted and connected graphs. Trans. Amer. Math. Soc. 78 (1955) 445-463.

HarF56   Harary, F. On the number of dissimilar line-subgraphs of a given graph. Pacific J. Math. 6 (1956).

HarF57   Harary, F. The number of oriented graphs, Michigan Math. J. 4 (1957) 221-224.

HarF57a   Harary, F. The number of dissimilar supergraphs of a linear graph. Pacific J. Math. 7 (1957) 903.

HarF58   Harary, F. On the number of bicolored graphs. Pacific J. Math. 8 (1958) 743-755.

HarF58a   Harary, F. On the number of dissimilar graphs between a given graph-subgraph pair. Canad. J. Math. 10 (1958) 513-516.

HarF59b   Harary, F. The number of functional digraphs. Math. Annalen. 138 (1959) 203-210.

HarF60   Harary, F. Unsolved problems in the enumeration of graphs. Proc. Math. Inst. Hung. Acad. Sci. Ser A 5 (1960) 63-95.

HarF64   Harary, F.   Combinatorial   problems   in   graphical
           enumeration. Applied Combinatorial Mathematics (E. F.
           Beckenbach, ed.) Wiley, New York (1964) 185-217.
HarF65   Harary, F., Palmer, E. M. The power group of two per-
           mutation groups. Proc. Nat. Acad. Sci. U.S.A. 54 (1965)
           680-682.
HarF66   Harary, F., Palmer, E. M. The power group enumeration
           theorem. J. Comb. Theory 1 (1966) 157-173.
HarF67   Harary, F. A proof of Polya's enumeration theorem. A
           seminar in graph theory. (F. Harary, ed.) Holt, Rinehart
           and Winston, New York (1967) 21-24.
HarF67a   Harary, F. Enumeration of graphs and digraphs. A
           seminar in graph theory. (F. Harary, ed.) Holt, Rinehart
           and Winston, New York (1967) 34-41.
HarF67b   Harary, F. Graphical enumeration problems. Graph
           Theory and Theoretical Physics (F. Harary, ed.) Academic
           Press, London (1967) 1-41.
HarF67c   Harary, F., Palmer, E. M. The enumeration methods of
           Redfield. Amer. J. Math. 89 (1967) 373-384.
HarF67d   Harary, F., Palmer, E. M. Enumeration of finite auto-
           mata. Inf. and Control 10 (1967) 499-508.
HarF70   Harary, F., Read, R. C. Enumeration of tree-like poly-
           hexes. Proc. Edin. Math. Soc. Ser. II 17 (1970) 1-13.
HarF70a   Harary, F. Enumeration under group action: Unsolved
           problems in graphical enumeration IV. J. Combinatorial
           Theory 8 (1970) 1-11; 9 (1970) 221.
HarF71   Harary, F. Graph Theory. Addison-Wesley Pub. Co., 1971.
HarF73   Harary, F., Palmer, E. M. Graphical Enumeration.
           Academic Press, 1973.
HarF75   Harary, F., Robinson, R. W. The number of achiral trees.
           J. Reine Angen. Math. (1975).
HarF75a   Harary, F., Palmer, E. M., Read, R. C. On the cell-growth
           problem for arbitrary polygons. Discrete Math. 11 (1975)
           371-389.
HarF75b   Harary, F. Twenty step algorithm for determining the
           asymptotic number of trees of various species. J. Austral.
           Math. Soc. 20 (1975) 483-503.
HarM63   Harrison, M. A. The number of classes of invertible
           Boolean functions. J. ACM 10 (1963) 25-28.
HarM63a   Harrison, M. A. The number of transitivity sets of
           Boolean functions. SIAM J. 11 (1963) 886-878.
HarM64   Harrison, M. A. On the classification of Boolean functions
           by the general linear and affine groups. SIAM J. 12
           (1964) 285-299.
HarM65   Harrison, M. A. A census of finite automata. Canad. J.
           Math. 17 (1965).

HarM71   Harrison, M. A. Counting theorems and their applications to classification of switching functions. Chap. 4 of Recent Developments in Switching Theory (A. Mukhopadyay, ed.) Academic Press (1971).

HarM73   Harrison, M. A. On the number of classes of binary matrices. IEEE Trans. Comput. C-22 12 (1973) 1048-1051.

HasW84   Hässelbarth, W. A note on Pólya's enumeration theorem. Theor. Chim. Acta 66 (1984) 91-110.

HerF82   Hering, F., Read, R. C., Shephard, G. C. The enumeration of stack polytopes and simplicial structures. Discrete Math. 46 (1982) 203-217.

HilT43   Hill, T. L. On the number of structural isomers in simple ring compounds - I. J. Phys. Chem. 47 (1943) 253-260.

HilT43a   Hill, T. L. On the number of structural isomers in simple ring compounds - II. J. Phys. Chem. 47 (1943) 413-421.

HilT43b   Hill, T. L. An application of group theory to isomerism in general. J. Chem. Phys. 11 (1943) 294-297.

JesJ73   Jesson, J. P., Meakin, P. Determination of mechanistic information for nuclear magnetic resonance line shape for intramolecular exchange. Accounts of Chem. Res. 6 (1973) 269-275.

JucA76   Jucys, A. - A.A. The number of distinct Latin squares as a group theoretic constant. J. Comb. Theory 20 (1976) 265-272.

KenB64   Kennedy, B. A., McQuarrie, D. A., Brubaker, C. H. Group theory and isomerism. Inorg. Chem. 3 (1964) 265-268.

KinC64   King, C., Palmer, E. M. Calculation of the number of graphs of order $p = 1(1)24$. Unpublished report (see [HarF73]).

KinR81   King, R. B. Chemical applications of topology and group theory. 9. The symmetries of coordination polyhedra. Inorg. Chem. 20 (1981) 363-372.

KleD75   Klein, D. J., Cowley, A. H. Permutational isomerism. J. Am. Chem. Soc. 97 (1975) 1633.

KleD81   Klein, D. J. Rigorous results for branched polymer models with excluded volume. J. Chem. Phys. 75 (1981) 5186-5189.

KleW72   Klemperer, W. G. Enumeration of permutational isomerization reactions. I. J. Chem. Phys. 56 (1972) 5478. II. Inorg. Chem. 11 (1972) 2668-2678.

KnoO75   Knop, O., Barker, W. W., White, P. S. Univalent (monodentate) substitution on convex polyhedra. Acta Cryst. 31 (1975) 461-472.

KonD36   König, D. Theorie der Graphen. Leipzig, 1936. Reprinted by Chelsea Publishing Company, New York, 1950.

KopF75   Kopp, F. O. die Ermittlung selbstanpressungsfähiger Zwischenräde-Anordnung für Reibrädergetreibe mit Hilfe von Graphen. Mechanism and Machine Theory 10 (1975) 521-529.

KorM75   Kornilov, M. Y.   Number of structural isomers in the
            adamantane series. J. Structural Chem. 16 (1975) 466-468.
            (Translated from Zh. Strukturnoi Khimii 16 (1975)
            495-498).

LarA68   Larsen, A. H., Pings, C. J. Counting graphs of interest in
            statistical mechanics including nonadditivity effects. J.
            Chem. Phys. 49 (1968) 72-80.

LeoJ75   Leonard, J. E., Hammond, G. S., Simons, H. E. The
            apparent symmetry of cyclohexane. J. Am. Chem. Soc. 97
            (1975) 5052-5054.

LeoJ77   Leonard, J. E.   Isomer numbers of nonrigid molecules.
            The cyclohexane case. J. Phys. Chem. 81 (1977) 2212-
            2214.

LisV69   Liskovec, V. A.   On a recurrence method of counting
            graphs with labelled vertices. Dokl. Akad. Nauk. SSSR 184
            (1969) No. 6 1284-1287. Translated in Soviet Math. Dokl.
            10 (1969) 242-246.

LitD50   Littlewood, D. E. Theory of Group Characters. Oxford
            (1950).

LiuC68   Liu, C. L.   Introduction to Combinatorial Mathematics.
            McGraw-Hill, New York 1968.

LiuC72   Liu, C. L.   Topics in combinatorial mathematics. MAA
            1972.

LloE68   Lloyd, E. K.   Pólya's theorem in combinatorial analysis
            applied to enumerate multiplicative partitions. J. London
            Math. Soc. 43 (1968) 224-230.

LloE84   Lloyd, E. K.   J. Howard Redfield 1879-1944. J. Graph
            Theory 8 (1984) 195-203.

LloE85   Lloyd, E. K.   Redfield's papers and their relevance to
            counting isomers and isomerizations. Unpublished article
            1985. To appear in Discrete Applied Mathematics.

LunA29   Lunn, A. C., Senior, J. K. Isomerism and configurations.
            J. Phys. Chem. 33 (1929) 1027-1079.

LupO59   Lupanov, O. B. Asymptotic estimates of the number of
            graphs having n branches. Dokl. Akad. Nauk. SSSR 126
            (1959) 498-500.

McLT78   McLarnen, T. J. The combinatorics of cation-deficient
            close-packed structures. J. Solid State Chem. 26 (1978)
            235-244.

McLT81   McLarnen, T. J.   Mathematics tools for counting poly-
            types. Z. fur Kristal. 155 (1981) 227-245.

McLT82   McLarnen, T. J., Baur, W. H. Enumeration of Wurtzite
            derivatives and related dipolar tetrahedral structures. J.
            Solid State Chem. 42 (1982) 283-299.

MasL74   Masinter, L. M., Sridharan, N. S., Lederberg, J., Smith, D.
            H. Applications of artificial intelligence for chemical
            inference. XII. Exhaustive generation of cyclic and
            acyclic isomers. J. Am. Chem. Soc. 96 (1974) 7702-7714.

MerR80 Merris, R. Manifestations of Pólya's counting theorem. Lin. Alg. and Appl. 32 (1980) 209-234.

MerR80a Merris, R. Pattern inventories associated with symmetry classes of tensors. Lin. Alg. and Appl. 29 (1980) 225-230.

MerR81 Merris, R. Pólya's counting theorem via tensors. Amer. Math. Monthly 88 (1981) 179-185.

MooP76 Moore, P. B., Asaki, T. The crystal structure of bredigite and the genealogy of some alkaline earth orthosilicates. American Mineralogist 61 (1976) 74-87.

MooP76a Moore, P. B., Asaki, T. Braunite: its structure and relationship to bixbyite and some insights on the genealogy of fluorite derivative structures. American Mineralogist 61 (1976) 1226-1240.

Mull69 Mullat, I. E. On the Redfield-Read combinatory algorithm. Mat. Zametki 6 (1969) 213-223. (Translated in Math. Notes 6 (1969) 583-588.)

MurF55 Murray, F. J. Mechanisms and robots. J. ACM 2 (1955) 61-82.

NeuP79 Neumann, P. A. A lemma that is not Burnside's. Math. Scientist 4 (1979) 133-141.

NouJ77 Nourse, J. G., Carhart, R. E., Smith, D. H., Djerassi, C. Generalized stereoisomerism modes. J. Amer. Chem. Soc. 99 (1977) 2063-2069.

NouJ79 Nourse, J. G. The configuration symmetry group and its application to stereoisomer generation, specification and enumeration. J. Am. Chem. Soc. 101 (1979) 1210-1216.

ObeW67 Oberschelp, W. Kombinatorische Anzahlbestimmungen in Relationen. Math. Ann. 174 (1967) 53-58.

OttR48 Otter, R. The number of trees. Ann. Math. 49 (1948) 583.

PalE70 Palmer, E. M. Asymptotic formulas for the number of self-complementary graphs and digraphs. Mathematika 17 (1970) 85-90.

PalE77 Palmer, E. M., Schwenk, A. J. The number of self-complementary achiral necklaces. J. Graph Theory 1 (1977) 309-315.

PalE85 Palmer, E. M. Graphical Evolution. An introduction to the theory of random graphs. Wiley-Interscience Series in Discrete Mathematics, John Wiley and Sons, 1985.

ParK68 Parthasarathy, K. R. Enumeration of graphs with given partition. Canad. J. Math. 20 (1968) 40-47.

PerD32 Perry, D. The number of structural isomers of certain homologs of methane and methanol. J. Amer. Chem. Soc. 54 (1932) 2918-2920.

PilR74 Pilgrim, R. L. C. The number of possible isomers in the porphyrins. J. Chem. Education 51 (1974) 316-318.

PolG35 Pólya, G. Un problème combinatoire général sur les groupes des permutations et le calcul du nombre des composés organiques. Comp. rend. Acad. Sci. Paris 201 (1935) 1167-1169.

PolG36   Pólya, G. Algebraische Berechnung der Isomeren einiger organischer Verbindungen. Z. für Krystallogr. (A) 93 (1936) 414.

PolG36a   Pólya, G. Über das Anwachsen der Isomerenzahlen in den homologen Reihe der organischen Chemie. Vierteljschr. Naturforsch. Ges. Zürich 81 (1936) 243-258.

PolG36b   Pólya, G. Sur le nombre des isomères de certains composés chimiques. Comp. rend. Acad. Sci. Paris 202 (1936) 1554.

PolG40   Pólya, G. Sur les types de propositions composées. J. Symbolic Logic 5 (1940) 98-103.

PreV76   Prelog, V. Chirality in chemistry. Science 193 (1976) 17-24.

QuiL77   Quintas, L. V., Yarmish, J. The number of chiral alkanes having carbon automorphism group isomorphic to a symmetric group. Proc. 2nd Caribbean Conf. in Comb. and Comput. Univ. of West Indies, Barbados (1977) 160-190.

QuiL79   Quintas, L. V., Yarmish, J. The number of chiral alkanes having given diameter and carbon automorphism group, a symmetric group. Proc. 2nd Internat. Conf. on Comb. Math., New York Acad. Sci. 1978 Annals of N.Y. Acad. Sci. vol. 319 (1979) 436-443.

ReaR58   Read, R. C. Some enumeration problems in graph theory. Doctoral Thesis, University of London, 1958.

ReaR59   Read, R. C. The enumeration of locally restricted graphs I. J. London Math. Soc. 34 (1959) 417-436.

ReaR60   Read, R. C. The enumeration of locally restricted graphs II. J. London Math. Soc. 35 (1960) 344-451.

ReaR61   Read, R. C. A note on the number of functional digraphs. Math. Annals 143 (1961) 109-110.

ReaR63   Read, R. C. On the number of self-complementary graphs and digraphs. J. London Math. Soc. 38 (1963) 99-104.

ReaR68   Read, R. C. The use of S-functions in combinatorial analysis. Canad. J. Math. 20 (1968) 808-841.

ReaR68a   Read, R. C. Some applications of a theorem of de Bruijn. Graph Theory and Theoretical Physics (F. Harary, ed.) Academic Press, 1968.

ReaR68b   Read, R. C. An introduction to chromatic polynomials. J. Comb. Theory 4 (1968) 52-71.

ReaR72   Read, R. C. Some recent results in chemical enumeration. Graph Theory and Applications (Y. Alavi, D. R. Lick, A. T. White, eds.) Lecture Notes in Math. 303 Springer 1972.

ReaR76   Read, R. C. The enumeration of acyclic chemical compounds. Chap. 4. Chemical Applications of Graph Theory (A. T. Balaban, ed.) Academic Press, 1976.

ReaR78   Read, R. C. On general dissections of a polygon. Aequat. Math. 18 (1978) 370-388.

ReaR80   Read, R. C., Wormald, N. C. Counting the 10-point graphs by partition. J. Graph Theory 5 (1981) 183-196.

RedJ27   Redfield, J. H. The theory of group reduced distributions. Amer. J. Math. 49 (1927) 433-455.

RedJ84   Redfield, J. H. Enumeration by frame group and range groups. J. Graph Theory 8 (1984) 205-224.

ReiD85   Reiner, D. L. Enumeration in music theory. Amer. Math. Monthly 92 (1985) 51-54.

RidR51   Riddell, R. J. Contributions to the theory of condensation. Ph.D. Dissertation. University of Michigan, 1951.

RidR53   Riddell, R. J., Uhlenbeck, G. E. On the theory of the virial development of the equation of state of monoatomic gases. J. Chem. Phys. 21 (1953) 2056.

RioJ58   Riordan, J. An introduction to combinatorial analysis. John Wiley & Sons 1958.

RobR70   Robinson, R. W. Enumeration of acyclic digraphs. Combinatorial Mathematics and Its Applications (R. C. Bose, et. al. eds.) Univ. of North Carolina, Chapel Hill (1970) 391-399.

RobR70a   Robinson, R. W. Enumeration of non-separable graphs. J. Comb. Theory 9 (1970) 327-356.

RobR73   Robinson, R. W. Counting labeled acyclic digraphs. New Directions in the Theory of Graphs (F. Harary, ed.) Academic Press (1973) 239-273.

RobR76   Robinson, R. W. Counting unlabeled acyclic digraphs. Lect. Notes in Math. 622 Springer (1976) 28-43.

RotG64   Rota, G-C. On the foundations of combinatorial theory. I. Theory of Möbius functions. Z. Wahrscheinlichkeitstheorie 2 (1964) 340-368.

RouD71   Rouvray, D. H. Graph theory in chemistry. R.I.C. Reviews 4, 2 (1971) 173-195.

RouD72   Rouvray, D. H. The mathematical theory of isomerism. Chemistry 45 (1972) 6-11.

RouD75   Rouvray, D. H. The pioneers of isomer enumeration. Endeavour 24 (1975) 28-33.

RouD75a Rouvray, D. H. Symmetry and isomer counts in the arenes. South African J. Science 71 (1975) 106-110.

RouD76   Rouvray, D. H., Balaban, A. T. Chemical applications of graph theory. Chap. 7 of Chemical Applications of Graph Theory (ed. A. T. Balaban) Academic Press, 1976.

RouD77   Rouvray, D. H. Sir Arthur Cayley - mathematician/chemist. Chemistry in Britain (1977) 52-57.

RucE70   Ruch, E., Hässelbarth, W., Richter, B. Doppelnebenklassen als Klassenbegriff und Nomenklaturprinzip für Isomere und ihre Abzählung. Theoret. Chim. Acta 19 (1970) 288-300.

RucE72   Ruch, E. Algebraic aspects of the chirality phenomenon in chemistry. Accounts of Chem. Res. 5 (1972) 49-56.

RucE83   Ruch, E., Klein, D. J. Double cosets in chemistry and physics. Theor. Chim. Acta 63 (1983) 447-472.

SheJ68   Sheehan, J. The number of graphs with a given automor-
          phism group. Canad. J. Math. 20 (1968) 1068-1076.
SlaZ81   Slanina, Z. Chemical isomerism and its contemporary
          theoretical description. Advances in Quantum Chemistry
          13 (1981) 89-153.
SleD53   Slepian, D. On the number of symmetry types of boolean
          functions of n variables. Canad. J. Math. 5 (1953) 185-193.
SolL77   Solomon, L. Partition identities and invariants of finite
          groups. J. Comb. Theory A 23 (1977) 148-175.
StaR73   Stanley, R. P. Acyclic orientations of graphs. Discrete
          Math. 5 (1973) 171-178.
StaR79   Stanley, R. P. Invariants of finite groups and their appli-
          cations to combinatorics, Bull. Amer. Math. Soc. New
          Series 1 (1979) 475-511.
SteM67   Stein, M. L., Stein, P. R. Enumeration of linear graphs
          and connected linear graphs up to $p = 18$ points. Report
          LA-3775 Los Alamos Scientific Lab. of the Univ. of
          California, Los Alamos, N.M. 1967.
StoP71   Stockmeyer, P. K. Enumeration of graphs with prescribed
          automorphism group. Ph.D. Thesis, U. of Michigan 1971.
TapR78   Tapscott, R. E. Enumeration of permutational isomers.
          The porphyrins. J. Chem. Educ. 55 (1978) 446-447.
TayW43   Taylor, W. J. Applications of Pólya's Theorem to opti-
          cal, geometrical and structural isomerism. J. Chem. Phys.
          11 (1943) 532.
TriN83   Trinajstić, N., Jericević, Z. Computer generation of iso-
          meric structures. Pure Appl. Chem. 55 (1983) 379-390.
TucA74   Tucker, A. Pólya's enumeration formula by example.
          Math. Mag. 47 (1974) 248-256.
WhiD74   White, D. E. Linear and multilinear aspects of isomorph
          rejection. Linear and multilinear algebra (1974) 211-226.
WhiD75   White, D. E. Classifying patterns by automorphism group:
          an operator theoretic approach. Discrete Math. 13 (1975)
          277-295.
WhiD75a   White, D. E. Multilinear enumerative techniques. Linear
          and Multilinear Algebra 4 (1975) 341-352.
WhiD75b   White, D. E. Redfield's theorems and multilinear algebra.
          Canad. J. Math. 27 (1975) 704-714.
WhiD75c   White, D. E. counting patterns with a given automor-
          phism group. Proc. Amer. Math. Soc. 47 (1975) 41-44.
WhiD76   White, D. E., Williamson, S. G. Combinatorial structures
          and group invariant partitions. Proc. A.M.S. 55 (1976) 233-
          236.
WhiD80   White, D. E. A Pólya interpretation of the Schur func-
          tion. J. Comb. Theory A 28 (1980) 272-281.
WilD78   Wille, D. Enumeration of self-complementary structures.
          J. Comb. Theory B 25 (1978) 143-150.

WilS70   Williamson, S. G. Operator theoretic invariants and the enumeration theorem of Pólya and de Bruijn. J. Comb. Theory 8 (1970) 162-169.

WilS71   Williamson, S. G. Pólya's counting theorem. J. London Math. Soc. 3 (1971) 411-421.

WilS72   Williamson, S. G. The combinatorial analysis of patterns and the principle of inclusion-exclusion. Discrete Math. 1 (1972) 357-388.

WilS73   Williamson, S. G. Isomorph rejection and a theorem of de Bruijn. SIAM J. Comput. 2 (1973) 44-59.

WilS73a   Williamson, S. G. Tensor compositions and lists of combinatorial structures. Linear and Multilinear Algebra 1 (1973) 119-138.

WilR76   Wilson, R. J. Graph theory and chemistry. Colloq. Math. Soc. Janos Bolyai (1976) 1147-1164.

WorN78   Some problems in the enumeration of labelled graphs. Ph.D. thesis. University of Newcastle, 1978.

WorN81   Wormald, N. C. On the number of planar maps. Canad. J. Math. 33 (1981) 1-11.

WriE67   Wright, E. M. A relationship between two sequences, I, II. Proc. London Math. Soc. 17 (1967) 296-304, 547-552.

WriE68   Wright, E. M. A relationship between two sequences, III. J. London Math. Soc. 43 (1968) 720-724.

WriE69   Wright, E. M. The number of graphs on many unlabelled nodes. Math. Annal. 183 (1969) 250-253.

WriE70   Wright, E. M. Asymptotic relations between enumerative functions in graph theory. Proc. London Math. Soc. 20 (1970) 558-572.

WriE71a   Wright, E. M. Graphs on unlabelled nodes with a given number of edges. Acta Math. 126 (1971) 1-9.

WriE72   Wright, E. M. The probability of connectedness of an unlabelled graph can be less for more edges. Proc. Am. Math. Soc. 35 (1972) 21-25.

WriE74   Wright, E. M. Asymmetric and symmetric graphs. Glasgow Math. J. 15 (1974) 69-73.

WriE74a   Wright, E. M. Graphs on unlabelled nodes with a large number of edges. Proc. London Math. Soc. 28 (1974) 577-594.

WriE81   Wright, E. M. Burnside's lemma. A historical note. J. Comb. Theory B (1981) 89-90.